America

From Zipper to SpaceX

Immigrant's Angle

America

From Zipper to SpaceX

Donald Sobot

WWW.AMERICAFROMZIPPERTOSPACEX.COM

DEDICATION

To the fearless dreamer and Harvard graduate, whose journey to America began with a heart brimming with ambition and a silent promise of greatness. You, who dared to dream beyond the horizons of your youth, stepping onto the most prestigious grounds of Harvard without whispering a word to those who held you dear, until your dream was firmly in grasp. Your journey mirrors the spirit of every dreamer who looks upon America as a land of endless possibilities. To you, whose name need not be spoken, who knows it's your story that inspired these pages - this book is dedicated.

TABLE OF CONTENTS

PART ONE

INTRODUCTION

HOW I FELL IN LOVE WITH AMERICA

You have probably heard many different stories about America, some good, some bad, some you understand, others you don't, some you don't have any idea about. Some stories come from very smart people, some from losers. Ordinary people, who live inside and outside of the USA, go about their everyday lives unaware of how deeply American innovations and culture have infiltrated their lives, taking these influences for granted. There are, also, those around the globe who love America and others who harbor negative sentiments toward America, some dislike America, some hate America, and few even wish to see America burn. Interestingly, most of these critics have never experienced the USA firsthand, and yet still are influenced by American innovation. I'm here to tell you about my firsthand observations of America and the American Dream, which have enlightened the whole world.

I arrived on American soil at the dawn of the 21st century and, in the time since, have become an American. This

American story is based on true, real facts and things that deeply affect all of us, both inside and outside of America. In the process of writing this book, I learned a lot about America, and I hope that by reading it, you too will broaden your understanding, perception, and perspective of America. By recognizing the small, everyday elements that enhance our lives, we can foster a deeper understanding for each other.

When I first set foot on American land, it was with a mix of awe and skepticism. America, the land renowned for its sky-scraping buildings, technological advancements, and a melting pot of cultures, was now the land I would call home. It was a nation built on the backbone of inventions, innovations, and ideas – some so monumental that they have shaped not just American life but have rippled across the world.

As I delved deeper into the American story, I began to realize that America's true essence isn't just in its global cultural influence through music, movies, or fashion. It's in these everyday inventions and innovations that have quietly woven themselves into the fabric of daily life across the globe. From the light bulb to the internet, from the automobile industry to the smartphone, America's contributions are omnipresent, yet often unacknowledged or taken for granted. These inventions have not just changed American lives; they have transformed the world in profound ways.

This realization was the genesis of my love for America. It was not just the grandeur of its cities or the beauty of its landscapes that captivated me, but the spirit of innovation and relentless pursuit of progress that seemed to be the nation's heartbeat. It was the stories of individuals, whose tireless work ethic and countless inventions exemplify the relentless American drive to push boundaries and create. Moreover,

America's influence extends beyond the tangible inventions. It's in the realm of ideas and values - the concept of a 'melting pot' society, where diversity is celebrated, and different cultures coexist and contribute to a richer, more varied national identity. It's in the American legal and political systems, whose concepts of democracy, rights, and freedoms have inspired similar systems and reforms around the world.

As I explored these facets of America, I understood that my story was also becoming part of the larger American narrative. The opportunities I found, the freedom to express myself, to innovate and to pursue my dreams, were facets of the American life that millions around the world aspire to. My journey was a testament to the American Dream, a concept that has become almost mythical in its promise of offering everyone, regardless of their origin, an equal chance at success and happiness. America's story is one of relentless innovation and transformation. It's a narrative punctuated by breakthroughs and setbacks, dreams, and disappointments. This journey, which stretches from the humble zipper to the ambitious endeavors of SpaceX, is emblematic of America's spirit. These stories are not just about technological advancements; they are narratives of perseverance, creativity, and the unyielding American spirit. They showcase a nation constantly striving to push the boundaries of what is possible, both for itself and for humanity at large.

As we journey through these chapters, we will see how American inventions and innovations have become intertwined with our daily lives, often in ways we hardly notice. They have not only shaped American society but have also had profound impacts on global culture, economics, and politics. My love for America grew as I delved into these stories, recognizing the country's role not just as a geopolitical entity but as a crucible of innovation and ideas. This book is an invitation to see America through this lens — to appreciate the ingenuity, resilience, and vision that have made America a land of endless possibilities and a beacon of hope and progress.

'America - From Zipper to SpaceX' is more than a chronicle of inventions and innovations; it's a narrative about a nation's journey, its impact on the world, and how, in the process of understanding America's story, we can come to understand a little more about ourselves and our place in this interconnected world. It's a story of how a nation's innovations can transcend borders, shaping not just a country but the fabric of global society. This narrative is a testament to the power of ideas and the impact of embracing change and pursuing progress. In each chapter, as we explore these inventions and the stories behind them, we not only trace the arc of American innovation but also understand how these inventions mirror the American ethos — an ethos characterized by a relentless pursuit of progress, a willingness to challenge the status quo, and a deep belief in the power of the individual.

Through this exploration, 'America - From Zipper to SpaceX' is not just a celebration of American ingenuity; it's an invitation to reflect on the broader implications of these inventions for our global community. It underscores the idea that innovation is not just about creating new technologies or products, but about inspiring change, fostering progress, and building a better future. As I conclude this introduction, I invite you to join me on this journey through the pages of history, to discover how America, with its tapestry of people, ideas, and inventions, has shaped and continues to shape the world. Let us embark on this exploration together, with open minds and a keen sense of curiosity, to appreciate the myriad ways in which American innovation has touched our lives and continues to chart the course of our collective future.

PART TWO

THE AMERICAN DREAM

"AMERICA IS LAST BEST HOPE"
-ABRAHAM LINCOLN-

In the heart of a bustling American city, under the shadow of gleaming skyscrapers and amidst the hum of relentless innovation, there lived a young inventor named Alex. Alex, inspired by the famous words of Abraham Lincoln, believed deeply in America as the "last best hope" of the world. His workshop was a microcosm of the nation's spirit: a place brimming with ideas and the relentless pursuit of making life better for everyone, everywhere.

Alex's journey began with his fascination for the great American inventors who had paved the way: from Thomas Edison's electric light bulb to the Wright brothers' airplane. Each story was a testament to the nation's ingenuity and its commitment to progress. In his small corner, Alex dreamt of contributing to this legacy.

His first invention was a simple yet revolutionary device that made clean water accessible in remote areas. It was a breakthrough that resonated with Lincoln's vision of America leading the world in making a difference. The device, powered by a unique combination of solar energy and nanotechnology, was a beacon of hope in places where clean water was a luxury.

But Alex didn't stop there. He knew that America's strength lay not just in groundbreaking inventions, but also in the myriad of small, everyday innovations that made life easier. He developed a range of products: a self-tilling garden tool that encouraged sustainable living, a compact and efficient battery that powered homes, and even an app that helped people manage their health better.

Each invention, though seemingly small, had a ripple effect. In villages across Africa, his water device brought life and health. In urban gardens, his tool empowered people to grow their own food. His battery innovation reduced the carbon footprint of homes, and his health app brought awareness and prevention to the forefront.

As Alex's inventions gained global recognition, he realized that his dream was aligning with the vision Lincoln once had. America, through its spirit of freedom, innovation, and a relentless desire to lead, was indeed making the world a better place. It wasn't just the big inventions that counted, but also the numerous small ones that touched everyday lives.

Years later, as an old man, Alex stood in his workshop, now a museum of American innovation. He gazed at the photos of Lincoln and other great inventors adorning the walls. With a content smile, he realized that the true essence of America's "last best hope" lay in the collective effort of its people to make the world a better, easier place for everyone. In this endeavor, every small invention mattered, each one a thread in the fabric of a better world.

Who's Alex?

Alex is a fictional character I created for this short story. He represents the spirit of American innovation and progress, inspired by Abraham Lincoln's quote, "America is the last best hope." The story uses Alex as a symbol to explore the impact of both major and minor American inventions on improving lives worldwide, embodying the ideals of freedom and leadership.

NONFICITION AMERICA

But now let's go to the nonfiction part of the famous Lincoln quote "America is last best hope". If we, for many different reasons, don't agree that America is the best place in the world, we should, at least, agree that America is **THE LEAST WORST PLACE IN THE WORLD**.

Correct?

In the last 100-150 years, America has been the place that has shaped the world with so many of their small and huge inventions that all people in the world have enjoyed and have made their lives much easier and more enjoyable.

Transitioning from the fictional narrative of Alex, let's delve into the non-fictional aspects of America's significant contributions to the world, particularly through its inventions and innovations over the past 100 to 150 years. This exploration aligns with the essence of Abraham Lincoln's quote, suggesting that while America may not be universally seen as the 'best' place, its impact through inventions and innovations positions it as a profoundly influential nation, arguably "the least worst place in the world" in terms of its global contributions.

Since the late 19th century, America has been the crucible of innovation. It began with Thomas Edison's invention of the electric light bulb in 1879, a breakthrough that not only revolutionized home and industrial lighting but also laid the foundation for the modern electric utility industry. This was followed by the Wright brothers' first powered flight in 1903, an event that fundamentally changed global transportation and connectivity.

Throughout the 20th century, American innovations continued to shape everyday life. In 1927, Philo Farnsworth developed the first electronic television, a device that transformed entertainment and media. The invention of the transistor in 1947 by William Shockley, John Bardeen, and Walter Brattain, marked a significant advancement in electronics, leading to the development of almost all modern electronic devices.

The 1960s and 1970s saw a series of space-related advancements, highlighted by the Apollo 11 moon landing in 1969, showcasing America's leading role in space exploration. This era also witnessed the emergence of the Internet, originally developed as ARPANET in 1969, which later became a global network connecting billions of people.

Medical advancements were not left behind. The development of vaccines for diseases like polio, spearheaded by American researchers like Jonas Salk, played a crucial role in global health. Technological innovations in the late 20th and early 21st centuries, including the personal computer, the mobile phone, and the development of the World Wide Web have further defined modern life.

In more recent years, American companies like Google, Apple, and Tesla have been at the forefront of innovations in software, hardware, and sustainable energy, respectively. The spirit of Silicon Valley, characterized by a culture of entrepreneurship and technological advancement, continues to influence global technology trends.

While America's contributions through inventions and innovations are indisputable, it's important to recognize that these advancements are part of a larger global narrative. Many of these inventions were built upon prior discoveries from around the world, and in turn, they have inspired further innovations globally. This interconnectivity highlights the global nature of scientific and technological progress, where each nation contributes to the collective advancement of humanity.

Reflecting on Lincoln's view of America as "the last best hope," it's evident that through its history of inventions and innovations, America has played a pivotal role in shaping the modern world. These contributions, both big and small, have had a profound and lasting impact on global society, making everyday life easier and more enjoyable for people around the world. Let's dive into some of them.

PART THREE

INVENTIONS & INNOVATIONS

AIRPLANE

At the end of the 19th century, in a land called America, a dream took flight. It was a dream to conquer the skies, to soar like birds, and to touch the clouds. This dream would change the world forever. In the late 1800s, two brave brothers from Ohio, Wilbur and Orville Wright, dared to dream this impossible dream. They believed that people could fly, not just in their wildest dreams, but in reality. They were about to make history.

American brothers, Wilbur and Orville Wright, were not scientists or engineers with fancy degrees. They were just ordinary bicycle makers with an extraordinary passion for aviation. They believed that if they could build a machine that could glide through the air, it would be like a bridge connecting people and places in ways never before imagined. The Wright brothers were not alone in their quest to conquer the skies, but they were different. They were determined to succeed when others gave up. They were pioneers, ready to explore the unknown.

In 1903, the Wright brothers packed up their flying machine and headed to Kitty Hawk, North Carolina. They chose this remote spot because of its strong winds, which they believed would help their flying machine take off. On December 17, 1903, the world held its breath as the Wright brothers prepared for their first flight. The airplane they had built was a fragile-looking contraption made of wood and fabric, with a small engine. It didn't look like much, but it was a marvel of engineering.

As the morning sun cast its golden rays over the sandy dunes of Kitty Hawk, the Wright brothers prepared for history. Wilbur climbed onto the machine while Orville stood by, ready to assist. The engine roared to life, and the aircraft moved down the rail they had set up for takeoff. Then, a miracle happened. The airplane lifted off the ground, defying gravity. For 12 seconds, it flew, covering a distance of 120 feet. The world had witnessed the first powered, controlled, and sustained flight of an aircraft. It was a moment that changed everything. The Wright brothers didn't stop at that one flight. They continued to refine and improve their flying machine and invented a system of control that allowed the pilot to steer the aircraft in three dimensions: up and down, left and right, and even turn. Their relentless dedication to innovation led to the development of more advanced airplanes. They demonstrated their flying skills to the world, and people were amazed.

The news of the Wright brothers' achievement spread like wildfire, not just in America but across the globe. People everywhere were inspired by their story. Aviation clubs and schools started popping up, and enthusiasts from different countries began building their own airplanes. Air travel was no longer just a dream. It was becoming a reality, thanks to the pioneering spirit of two ordinary Americans who dared to dream big. Today, when you board an airplane to visit friends, family, or explore new places, remember that it all began with a dream in America. The Wright brothers' legacy lives on in every flight you take. They didn't just invent an airplane; they gave the world a gift that connects us all.

From the Wright brothers' humble beginnings in Ohio to the skies above every country on Earth, their dream has truly taken flight. So, the next time you look up at an airplane soaring overhead, remember that it was American ingenuity and determination that made it possible for people all over the world to fly. The Wright brothers' achievement had a profound impact on society. Before their invention, travel was limited to the ground and sea. Air travel opened up new possibilities, shrinking the world and bringing people closer together. It revolutionized commerce, tourism, and communication. Businesses could now reach markets in distant places faster than ever before. Families separated by long distances could reunite in a matter of hours, not weeks. And the exchange of ideas and cultures became more accessible, fostering greater understanding among nations.

As the aviation industry grew, so did the demand for skilled pilots, engineers, and mechanics. Aviation schools and training programs sprang up across the United States and around the world. The field of aeronautics expanded rapidly, leading to innovations in aircraft design, navigation, and safety. The development of commercial aviation allowed people to travel for work and leisure, further connecting communities and nations. Airports became bustling hubs of activity, serving as gateways to the world.

The spirit of exploration and adventure that the Wright brothers embodied continued to inspire generations of aviators. Pioneers like Charles Lindbergh, who made the first solo nonstop transatlantic flight, and Amelia Earhart, who broke barriers for women in aviation, followed in their footsteps. Space exploration, which began with the historic Apollo 11 moon landing in 1969, was also fueled by the same human curiosity and determination that led to the first flight at Kitty Hawk. The legacy of the Wright brothers extended beyond Earth's atmosphere.

In a world filled with differences and divisions, aviation became a symbol of unity. Airplanes carried passengers and cargo across borders, transcending political boundaries. They brought aid to those in need during times of crisis and disaster, demonstrating the power of cooperation among nations. The

Wright brothers' dream of connecting people and places had come true on a global scale. No longer were oceans and mountains insurmountable barriers. The world had become a smaller, more interconnected place, all thanks to the innovation born in America.

As we stand on the threshold of the future, the legacy of Wilbur and Orville Wright continues to inspire. Aircraft have evolved into marvels of technology, capable of carrying hundreds of passengers at incredible speeds. Research into cleaner and more sustainable aviation solutions is ongoing, as we strive to protect the environment while maintaining the gift of flight.

The dream that began in the hearts of two brothers over a century ago lives on in every takeoff and landing. That's the incredible story of how the first airplane was invented and developed in the USA, changing the world and bringing people closer together through the gift of flight. From a humble wooden aircraft at Kitty Hawk to the bustling airports of today, the dream of flight has transformed our world, connecting people, cultures, and nations in ways unimaginable to those two bicycle makers from Ohio. The sky is no longer the limit; it's just the beginning of our journey. The dream lives in the hearts of those who continue to push the boundaries of what's possible in the skies... drones, fully automated pilots, space travel with SpaceX, etc. The future will tell us what is next.

LIGHT BULB

**Illuminating Innovation: Thomas Edison and the
Invention of the Light Bulb**

The Darkness of Night

Imagine a world without electric light, where darkness ruled the night. Before the electric light bulb, people relied on candles, oil lamps, and gas lighting for artificial illumination. These methods were expensive, produced limited light, and posed significant fire hazards.

The story of the electric light bulb begins with Thomas Edison, an American inventor known for his prolific contributions to technology. Edison was not the first to experiment with electric lighting, but he was determined to create a practical, commercially viable solution. One of the primary challenges Edison faced was finding a suitable filament that could glow brightly without burning out quickly. He tested thousands of materials and designs, eventually settling on a carbonized bamboo filament. Edison established his famous Menlo Park laboratory in New Jersey, often referred to as the "invention factory." It was here that he and his team worked tirelessly to develop and perfect the electric light bulb.

In 1879, Edison achieved a major breakthrough when he successfully tested a practical incandescent light bulb that could provide reliable, long-lasting illumination. This marked a turning point in the history of artificial lighting. Edison's invention attracted investors, and in 1878, he founded the Edison Electric Light Company to develop and commercialize his electric lighting system. This laid the foundation for the widespread adoption of electric lighting. Edison's electric light bulbs quickly found their way into homes, businesses, and streets. The installation of electric streetlights in cities like New York and London marked the beginning of safer and more efficient urban lighting.

In a world without electric light people would be limited to daylight hours for work and leisure. The absence of electric light would hinder progress in industry, transportation, and quality of life. Electric light had a profound impact on urban development. Cities could now function around the clock, stimulating economic growth, entertainment, and nightlife. The nighttime cityscape transformed with illuminated signs and landmarks. Thomas Edison's contributions extended beyond the light bulb. He was a driving force behind the development of the electrical power industry, including the creation of power plants and the distribution of electricity.

The invention of the electric light bulb exemplifies American innovation and its impact on the world. Edison's work influenced industries, changed daily routines, and shaped the modern world. Edison's light bulb is not just an invention; it's a symbol of human ingenuity and the power of innovation to overcome challenges. It remains a fundamental part of our daily lives. As we look to the future, lighting technology continues to evolve. Energy-efficient LED lighting has become the new standard, reducing electricity consumption and environmental impact.

Edison's work not only led to technological advancements but also fostered a culture of entrepreneurship and innovation in the United States. It inspired countless inventors and innovators to pursue their ideas and change the world. The invention of the electric light bulb by Thomas Edison is a testament to American ingenuity and its profound impact on the world. It brought light to the darkness, transformed cities, and ushered in a new era of convenience and productivity. As we celebrate the impact of this invention on our global society, we must also recognize the enduring legacy of innovation and the potential for even greater advancements in the future.

TELEPHONE

The Invention of the Telephone

In the late 19th century, amidst the whirlwind of technological innovation, there emerged an American invention that would change the way the world communicated—the telephone. This chapter explores the visionary minds behind the telephone's invention, emphasizing its American origins and its profound impact on people worldwide.

The tale of the telephone begins with Alexander Graham Bell, a Scottish-born inventor who had settled in the United States. Bell, along with his collaborators, Elisha Gray and Antonio Meucci, played crucial roles in the development of this groundbreaking communication device. In 1876, Alexander Graham Bell received a patent for his invention of the telephone. It was a momentous achievement that marked the birth of a new era in communication. Bell's telephone could transmit sound over distances, allowing people to speak to each other miles apart, a feat previously thought impossible.

The world's first telephone call took place on March 10, 1876, when Bell spoke to his assistant, Thomas Watson, saying, "Mr. Watson, come here, I want to see you." Those words, transmitted via telephone, were a historic milestone, marking the first time that human speech had been carried by electrical wires. While the telegraph had already enabled long-distance communication, it was the telephone that brought real-time, voice-based communication to the masses. The telephone quickly surpassed the telegraph in popularity, offering a more personal and immediate way to connect with others. The impact was swift and far-reaching. Bell's American Telephone and Telegraph Company (AT&T) played a pivotal role in the telephone's expansion, constructing a vast network of telephone lines across the United States. The telephone soon connected cities, towns, and rural areas, bringing people closer together.

The telephone was not confined to American soil; it quickly spread worldwide. Inventors, entrepreneurs, and governments in other countries recognized its potential and worked to establish their own telephone networks. The telephone became a symbol of progress and modernity, bridging gaps and fostering connections across borders. It was a game-changer for businesses. It facilitated faster communication, enabling companies to conduct transactions and negotiations more efficiently. It revolutionized customer service, allowing businesses to interact with customers in real time. The telephone transformed daily life. It brought families closer, enabling them to stay connected across great distances. It allowed for quick coordination, reducing the need for physical presence. It became an essential tool for emergency communication, helping save lives in times of crisis. The telephone became a symbol of progress and a muse for artists, writers, and musicians. It featured prominently in literature and cinema, serving as a plot device that drove stories forward. It also inspired art, including paintings and sculptures that captured its significance.

The telephone played a crucial role during times of crisis, from natural disasters to wartime. It enabled rapid communication for emergency response and coordination. During World War II, it was instrumental in military communications and intelligence. The telephone didn't stop evolving with its invention. Advancements in technology led to innovations like rotary dial phones, push-button phones, and cordless phones. These improvements made telephony more user-friendly and accessible. In the late 20th century, mobile phones emerged as the next frontier in communication. Initially, they were large and clunky, but they soon became sleek, portable devices that allowed people to communicate on the go. The mobile phone revolutionized the way we connect, enabling us to be reachable anywhere, anytime.

The integration of the telephone with the internet and the development of smartphones ushered in a new era of communication. Smartphones, like the iPhone, combined telephony, computing power, and internet access in a single device, changing how we work, socialize, and access information. The invention of the telephone, pioneered by Alexander Graham Bell and his American collaborators, remains an enduring symbol of American innovation and ingenuity. It represents the spirit of exploration, discovery, and the relentless pursuit of a better way to communicate.

As we look ahead, the telephone's legacy lives on in the digital age. We continue to find new ways to connect, from video calls to instant messaging. The principles of instant, long-distance communication that the telephone introduced continue to shape our world. The telephone, born from the brilliant minds of American inventors, reshaped the way humanity communicates. From Alexander Graham Bell's first words transmitted through wires to the smartphone era of today, it has been a journey of innovation and progress.

This invention, rooted in American genius and determination, became a global phenomenon that transcended borders and cultures. It brought people closer together, transformed businesses, and served as a symbol of human ingenuity. As we navigate the ever-changing landscape of communication, we must remember that the telephone was the catalyst for this transformation, forever etching its place in history as an invention that connected the world and changed the course of human interaction.

ELECTRICITY & ALTERNATING CURRENT

Electrifying the World: Nikola Tesla and the Rise of
Alternating Current

Imagine a world without the simple flip of a switch, where darkness ruled the night and technology remained stagnant. This chapter delves into the electrifying story of how an American inventor named Nikola Tesla and the development of alternating current (AC) power systems forever changed the course of history. In the late 19th century, electricity was still in its infancy. Thomas Edison, another American inventor, had made strides with direct current (DC) power, but it had limitations. DC power could only travel short distances and wasn't suitable for widespread use. The world needed a better solution. Enter Nikola Tesla, a Serbian immigrant to the United States, who possessed a brilliant mind and an unquenchable curiosity. Tesla was fascinated by electricity and had a vision for a more efficient and powerful electrical system. He believed that alternating current was the key.

Tesla's vision clashed with Edison's DC system, leading to what historians now call the "War of Currents." Edison, with his established reputation, championed DC power, while Tesla championed AC power. It was a battle for the future of electricity. Tesla's AC system had a significant advantage over DC: it could transmit electricity over long distances with minimal loss of power. Tesla's AC power system laid the foundation for the development of the electrical grid we know today. It allowed for the efficient generation, transmission, and distribution of electricity across vast regions, bringing power to homes, businesses, and industries. But Tesla's contributions extended far beyond AC power. Nikola Tesla was a scientific visionary, ahead of his time in many ways. He not only revolutionized the field of electrical engineering with AC power but also had numerous other futuristic ideas and inventions that continue to inspire and impact our world today.

One of Tesla's most visionary ideas was wireless

transmission of power. He envisioned a world where electricity could be transmitted through the air, without the need for wires or cables. While this concept was met with skepticism during his time, modern advancements in wireless charging and energy transmission have made Tesla's dream a reality to some extent. Technologies like wireless charging pads for smartphones and electric vehicle charging without cords owe a debt to Tesla's pioneering work in this area.

Tesla also had a keen interest in renewable energy sources. Long before solar panels and wind turbines became mainstream, he was exploring ways to harness the power of the sun and wind to generate electricity. He even proposed the idea of a "world system" that could collect energy from the atmosphere and distribute it wirelessly. While this concept remains largely theoretical, it foreshadowed the growing importance of renewable energy in our quest for a sustainable future. Furthermore, Tesla was fascinated by the concept of robotics and automation. He designed and built a remote-controlled boat and envisioned a future where machines would carry out tasks remotely, making human labor more efficient and less dangerous. Today, we see his ideas materializing in industries like manufacturing, agriculture, and even space exploration, where robotic technology plays a crucial role.

Tesla's futuristic inventions weren't limited to electricity and automation. He also explored the possibilities of radio waves for communication and even developed the concept of a "teleforce" weapon, which foreshadowed the development of directed-energy weapons in modern warfare. In many ways, Nikola Tesla was a man ahead of his time, with ideas and inventions that continue to shape our world. His contributions extended beyond AC power, encompassing wireless energy transmission, renewable energy, robotics, and communication technologies. While some of his more ambitious ideas remain unrealized, they continue to inspire scientists, engineers, and innovators worldwide.

As Tesla's visionary ideas and inventions continue to find practical applications in our modern world, we are reminded of the profound impact he had on technological innovation. Without his relentless pursuit of scientific advancement and his willingness to push the boundaries of what was possible, our world would be vastly different. Imagine a world without the influence of Nikola Tesla's ideas. There would be no wireless communication, no renewable energy solutions, and no automated industries. The world would lack the visionary spark that has driven progress and innovation for over a century.

The story of Nikola Tesla and the development of alternating current power systems is a testament to the power of vision, determination, and American innovation. But it is also a story of a man whose scientific visionary ideas and inventions extended far beyond electricity and continue to shape our world today. Tesla's legacy is not confined to a single invention or discovery; it is a tapestry of innovation that weaves through the fabric of modern civilization. He saw beyond the limitations of his era and envisioned a future that was electrifyingly different from the world he knew. His futuristic ideas and inventions were not only groundbreaking but also far-reaching.

For instance, Tesla's fascination with wireless power transmission, although not fully realized in his time, laid the foundation for technologies like wireless charging we use today. Imagine the convenience of charging your devices without needing to plug them in. This convenience, rooted in

Tesla's visionary thinking, has transformed how we interact with technology. Tesla's commitment to harnessing renewable energy sources also remains a cornerstone of our sustainable future. In an age when fossil fuels still dominated, he foresaw the potential of solar and wind power. Today, we see the fruits of his foresight in the proliferation of solar panels and wind turbines as clean energy sources, reducing our carbon footprint and mitigating climate change.

His fascination with automation and robotics, demonstrated in his remote-controlled boat and automation ideas, foreshadowed the rise of industries where machines perform tasks with precision, efficiency, and safety that would be otherwise unattainable. From manufacturing plants to space exploration, Tesla's influence can be seen in the relentless pursuit of automation. Tesla's contributions extended even to the realm of communication. His experiments with radio waves laid the groundwork for the development of modern telecommunications systems that connect us across the globe. In an era where information flows freely and instantaneously, we owe a debt of gratitude to Tesla's pioneering work. Nikola Tesla was a scientist, engineer, and inventor whose genius transcended his time. He wasn't content with the status quo; he dared to dream beyond the horizon of possibilities. His futuristic ideas and inventions were not only remarkable in themselves but also served as a beacon for future generations of innovators.

In today's world, where technology evolves at an astonishing pace, we continue to draw inspiration from Tesla's legacy. As we push the boundaries of what's possible in fields like renewable energy, wireless communication, automation, and beyond, we stand on the shoulders of this scientific visionary. Nikola Tesla's contributions are not confined to the past; they are the seeds of future innovations and breakthroughs. In a world where we face unprecedented challenges, from climate change to the complexities of modern warfare, Tesla's ability to envision solutions beyond the grasp of his contemporaries serves as a reminder of the power of human imagination and determination. The story of Nikola Tesla and the development of alternating current power systems is a testament to the power of vision, determination, and American innovation. Still, it is also a story of a man whose scientific visionary ideas and inventions have not only shaped our past but continue to illuminate the path towards the future.

TELEVISION

Illuminating the World: The Invention of the Television

Once upon a time, in the land of innovation known as America, a vision began to take shape—a vision of transmitting moving images through the airwaves, into our homes. It was a dream that would lead to the invention of the television, changing the way we see the world forever.

The story of television begins with American inventors who saw the potential to bring images to life in our living rooms. Charles Francis Jenkins, an inventor from Washington, D.C., and Philo Farnsworth, a young genius from Utah, are among the key figures who played pivotal roles in this revolution. Jenkins was the first to demonstrate a mechanical television system in 1923, laying the groundwork for the technology that would soon become a household staple. Meanwhile, Philo Farnsworth, a brilliant and self-taught engineer, was working on an entirely new approach to television—electronic scanning. Little did they know that their work would shape the future of entertainment.

In 1928, Charles Jenkins unveiled his Radiomovie, a mechanical television system that allowed people to watch moving images transmitted through the airwaves. It was a marvel of engineering and ingenuity. People gathered around these early television sets to catch a glimpse of the future. While Jenkins was making strides with mechanical television, young Philo Farnsworth was quietly working on electronic television. In 1927, at the age of just 21, Farnsworth achieved a breakthrough—he transmitted the first electronic television image, a simple line, using his invention known as the Image Dissector. This marked the birth of electronic television, a technology that would eventually surpass the limitations of mechanical systems. In 1930, Philo Farnsworth conducted the first public demonstration of electronic television. He transmitted an image of a dollar sign, making it clear that his invention was capable of transmitting not only simple lines but detailed pictures.

The world watched in awe as Farnsworth's electronic television system showed its potential. It was a momentous occasion, a glimpse into the future of entertainment and information. While American inventors were making strides in television technology, it was a visionary businessman, David Sarnoff, who recognized its commercial potential. Sarnoff, the head of Radio Corporation of America (RCA), was determined to bring television into people's homes. RCA played a significant role in advancing television technology and standards. In 1939, they introduced the RCA TRK-12, one of the first commercially available televisions. It was a massive console, but it was a symbol of the future. In 1939, RCA launched the first television network in the United States, NBC. This marked the beginning of regular television broadcasts. The world was introduced to a new form of entertainment—one that combined sound and moving images, captivating audiences like never before.

Television quickly became a centerpiece of American households, changing the way families spent their evenings. It provided entertainment, education, and a window into the world. News, sports, and cultural events were now accessible to millions. The 1950s and 1960s are often referred to as the "Golden Age of Television." It was a time when iconic shows like "I Love Lucy," "The Twilight Zone," and "The Ed Sullivan Show" graced the small screen. Television became a powerful cultural force, shaping American society and influencing the world.

Television continued to evolve. In the 1950s, color television was introduced, adding a vibrant dimension to the viewing experience. In the decades that followed, television technology advanced rapidly, from remote controls to flat-screen displays. Television was not confined to American borders. Its influence spread worldwide, reaching remote villages and bustling cities. It became a vehicle for cultural exchange, enabling people to learn about different cultures and traditions.

Television played a crucial role in the space age. Millions watched in awe as the Apollo 11 mission landed on the moon in 1969. It was a momentous event, and television brought it live into homes across the globe, connecting humanity in a way never imagined. Television news became a powerful tool for informing the public about current events. From the Kennedy assassination to the fall of the Berlin Wall, television brought the world's most significant moments into our living rooms, making us witnesses to history. Cable television brought an even wider range of channels and programming choices to viewers. The 1980s and 1990s saw the proliferation of cable networks, catering to niche audiences and interests.

The internet revolutionized television once again. Streaming services like Netflix, Amazon Prime, and Hulu brought on-demand content to viewers, changing how we consume television. Viewers now have the power to choose what they watch and when they watch it. Television has left an indelible mark on culture. It has inspired fashion trends, iconic catchphrases, and even influenced political discourse. Shows like "Star Trek" have inspired technological innovation, while sitcoms like "Friends" have become cultural touchstones.

Television, born out of the vision and innovation of American inventors and entrepreneurs, has become a global phenomenon. Its impact on society, culture, and technology cannot be overstated. From the early mechanical marvels to today's streaming services, television continues to evolve, reflecting and shaping our world. As we stand on the brink of the future, television is undergoing another transformation. High-definition and 4K displays provide stunning visuals. Virtual reality and augmented reality promise immersive viewing experiences. Television is no longer confined to the living room; it's in our pockets, on our tablets, and all around us.

And that is the remarkable story of how television was invented in America, changed the world, and brought the magic of moving images into our lives. From humble beginnings to the global phenomenon, it is today, television has been a testament to human creativity and innovation. It all started with a vision, a dream, and a few ingenious Americans who believed in the power of transmitting images through the airwaves. Through their efforts, they illuminated our lives, bringing entertainment, information, and a window to the world right into our homes.

As we look to the future, we can only imagine what the next chapter of television will bring. One thing is certain: the spirit of innovation that brought television to life will continue to shape its evolution, and it will remain a beloved and influential part of our lives for generations to come.

INTERNET

The Invention of the Internet – Connecting the World

In the not-so-distant past, the world was a different place. Communication was slow, information was limited, and the idea of a global network of connected devices was nothing more than a dream. But that dream was about to become a reality, thanks to the inventive spirit of Americans. The story of the internet begins with a group of American pioneers who saw the potential of connecting computers together. In the 1960s, scientists and engineers like J.C.R. Licklider, Leonard Kleinrock, and Lawrence Roberts began laying the foundation for what would become the World Wide Web. These visionaries believed that computers could be linked to share information and resources, making it easier for people to collaborate and access knowledge. Their ideas laid the groundwork for the digital revolution.

In 1969, the first successful message was sent over ARPANET, a project funded by the U.S. Department of Defense. This event marked the birth of the internet as we know it today. ARPANET was designed to withstand a nuclear attack by decentralizing its communication network, a feature that would later prove crucial in its expansion. Email, one of the internet's most essential tools, was also an American invention. Ray Tomlinson, an engineer working on ARPANET, sent the first email in 1971. This simple act paved the way for global communication, allowing people to exchange messages instantaneously across the world.

While the internet was becoming a reality, the World Wide Web, the user-friendly interface that would make it accessible to everyone, was still a few years away. In 1989, British computer scientist Tim Berners-Lee, working at CERN in Switzerland, proposed a system for sharing information on the internet. However, it was American companies and institutions that helped bring the web to life. In 1993, a small

American software company called Netscape Communications released the Netscape Navigator web browser. This user-friendly browser made it easy for people to explore the web and paved the way for the explosive growth of the internet.

The mid-1990s saw the emergence of countless internet startups, many of them based in the United States. Companies like Amazon, eBay, and Google were founded during this era, forever changing how we shop, trade, and search for information. The dot-com boom was a testament to the potential of the internet and its ability to transform industries. It wasn't just about business; it was about creating a global community. Google, founded by Larry Page and Sergey Brin in 1998, revolutionized how we find information on the web. Their search engine made it possible to access the vast knowledge of the internet with just a few keystrokes. Google's impact was felt worldwide, becoming synonymous with online search.

In the early 2000s, American social media platforms like Facebook, Twitter, and YouTube changed the way people connect and share their lives. These platforms gave a voice to billions and enabled global conversations on a scale never seen before. The internet has transcended borders, connecting people from all corners of the globe. It's become an essential tool for business, education, healthcare, and more. During times of crisis, it has enabled remote work, learning, and communication, proving its resilience and adaptability.

As the internet has grown, so have its challenges. Issues like privacy, security, and misinformation have arisen. But Americans have continued to innovate, developing solutions to address these concerns. Today, the internet is more integral to our lives than ever before. With the advent of 5G technology, the internet's potential for connectivity and innovation is boundless. It's not just about connecting computers; it's about connecting everything, from smart cities to wearable devices.

The internet's global nature has raised questions about governance and regulation. International organizations and governments collaborate to ensure the internet remains open, secure, and accessible to all. The United States, as a pioneer in internet development, has played a key role in shaping internet governance. The internet has transformed society in profound ways. It has democratized information, empowered individuals and communities, and given voice to marginalized groups. It has also accelerated the pace of innovation, sparking advances in fields such as medicine, education, and entertainment.

The concept of the Internet of Things (IoT) envisions a world where everyday objects are connected to the internet, enabling them to collect and exchange data. IoT has the potential to revolutionize industries, from healthcare to agriculture, making processes more efficient and responsive to our needs. The internet has fueled economic growth, creating jobs, driving innovation, and enabling new business models. E-commerce, online advertising, and the gig economy are just a few examples of how the internet has reshaped the economy.

The internet has expanded access to education, offering online courses and resources to learners worldwide. It has also transformed traditional classrooms, enabling personalized learning experiences and fostering global collaboration among students and educators. During emergencies and natural disasters, the internet has played a crucial role in providing real-time information, coordinating relief efforts, and connecting affected communities. It has become a lifeline in times of crisis, enabling rapid response and recovery.

Despite its global reach, the internet has not reached everyone equally. The digital divide remains a challenge, with disparities in access and connectivity. Efforts are underway to ensure that the benefits of the internet are accessible to all, regardless of location or socioeconomic status. Starlink, a SpaceX program founded by Elon Musk, could be the solution to the world's connectivity issue. All people, regardless of location, could be connected to each other through satellite internet in the very near future.

The invention of the internet, driven by American visionaries and innovators, has transformed the world. It has shattered barriers, brought people together, and revolutionized how we live, work, and communicate. From the early days of ARPANET to the global phenomenon of the World Wide Web, the internet is a testament to the power of human ingenuity. It's a tool that empowers individuals, businesses, and nations to thrive in an interconnected world.

As we look to the future, the legacy of the internet reminds us that innovation knows no bounds. It's a reminder that even the wildest dreams can become a reality, and that the power to connect and create lies at our fingertips, thanks to American inventors who changed the world for all people around the world.

PC COMPUTER

Revolutionizing Computing: The American Invention of
the Personal Computer

The Birth of Apple & The Rise of Microsoft

Before the advent of personal computers, computing was largely confined to large mainframe systems used by government agencies, research institutions, and corporations. These computers were massive, expensive, and inaccessible to the average person. The story of the personal computer begins with American innovators who sought to make computing accessible to individuals. Pioneers like Steve Jobs, Steve Wozniak, and Bill Gates played pivotal roles in the development of the personal computer.

In 1976, Steve Jobs and Steve Wozniak co-founded Apple Computer, introducing the Apple I—a bare circuit board that laid the foundation for personal computing. Their vision was to create user-friendly machines that could be used by anyone, not just computer experts. Around the same time, Bill Gates and Paul Allen founded Microsoft, focusing on developing software for personal computers. Their breakthrough product, MS-DOS, became the operating system of choice for IBM's first personal computer, the IBM PC. In 1981, IBM released the IBM PC, a pivotal moment in the history of personal computing. What set it apart was its "open architecture," which allowed third-party manufacturers to create hardware and software that could be used with the IBM PC, fostering a thriving ecosystem.

The IBM PC's open architecture ignited the PC revolution. A multitude of hardware and software options emerged, creating a competitive market that drove innovation and brought down costs. The PC became a household and business staple. One of the most significant innovations in personal computing was the graphical user interface (GUI). Apple's Lisa and Macintosh computers introduced a user-friendly interface with icons, windows, and a mouse, making computing more intuitive and accessible.

Imagine a world without personal computers. Work and communication would rely heavily on paper-based processes. Access to information and entertainment would be limited. The absence of PCs would hinder progress in science, business, and education. Personal computers revolutionized education by providing students with access to information, interactive learning experiences, and creative tools. They transformed the way students learn and educators teach. Personal computers reshaped the business world by improving productivity, data management, and communication. They automated tasks, reduced paperwork, and introduced new business models, such as e-commerce and remote work.

The personal computer is a symbol of American innovation and entrepreneurship. American tech companies, like Apple and Microsoft, became global leaders in the industry, shaping the digital landscape and influencing societies worldwide. As we look to the future, personal computing will continue to evolve. Advancements in artificial intelligence, cloud computing, and quantum computing promise to revolutionize how we work, solve complex problems, and interact with technology. The invention of the personal computer, driven by American innovation, has transformed the world. It has democratized access to information, accelerated scientific discovery, and empowered individuals to create, communicate, and innovate.

The personal computer, exemplified by companies like Apple and Microsoft, represents the power of American ingenuity to reshape the way the world computes and interacts with technology. It has become an indispensable tool that has enriched lives, driven progress, and connected people across the globe. As we celebrate the impact of the personal computer on our global society, we must also acknowledge the responsibility that comes with this technology. It is up to us to ensure that the potential for positive change is realized, and that technology continues to be a force for innovation, empowerment, and progress.

STREAMING PLATFORMS

The genesis of internet streaming can be traced back to the early 1990s in America, a period marked by rapid technological advancements and the dawn of the digital age. The development of the Internet and digital compression technologies laid the groundwork for what would become internet streaming. One of the earliest forays into internet streaming was made by an American company, RealNetworks. Founded in 1994 by Rob Glaser, a former Microsoft executive, RealNetworks released RealAudio, a groundbreaking audio streaming technology, and later RealVideo for video streaming. These technologies were among the first to allow users to stream media over the Internet, heralding a new era in digital media consumption.

Netflix, founded in 1997 by Reed Hastings and Marc Randolph in Scotts Valley, California, began as a DVD rental service. However, it was its transition to online streaming in 2007 that marked a turning point in the history of media consumption. Netflix's streaming service allowed viewers to watch a wide array of TV shows and movies on demand, fundamentally changing how people accessed entertainment. Netflix's model of content delivery and its investment in original content transformed the entertainment industry. The idea of 'binge-watching' became synonymous with the Netflix experience, and the company's success prompted a shift in the industry towards streaming models.

In 2005, another monumental event in the history of internet streaming occurred. YouTube, created by Steve Chen, Chad Hurley, and Jawed Karim in San Bruno, California, was launched. YouTube's platform, which allowed users to easily upload and share videos, democratized video content creation and distribution. YouTube quickly became a cultural phenomenon. It provided a platform for individuals to share their lives, talents, and opinions with a global audience, disrupting traditional media and entertainment paradigms. The platform's impact extended beyond entertainment, influencing politics, education, and social movements.

Following the success of Netflix and YouTube, numerous other streaming services emerged. Hulu, launched in 2007 as a joint venture by several major media companies, focused on streaming TV shows and movies. Amazon Prime Video, part of Amazon's broader Prime subscription service, expanded into streaming, offering a diverse range of content, including critically acclaimed original series. These platforms not only competed for viewership but also drove innovation in content delivery, user experience, and business models, further cementing streaming as the future of media consumption.

The rise of streaming services had a profound impact on traditional media. Cable television and movie theaters faced declining audiences as consumers increasingly preferred the convenience and variety of streaming services. Moreover, streaming platforms began to invest heavily in original content, attracting top talent and competing with traditional media companies for viewership. The globalization of content is another significant aspect of the streaming revolution. Platforms like Netflix and Amazon Prime Video made international content accessible to global audiences, breaking down geographic and cultural barriers in entertainment. Advancements in technology have continuously shaped the streaming landscape. Improvements in internet speed and bandwidth, along with the proliferation of mobile devices, have made streaming more accessible than ever before. The development of smart TVs and devices like Roku and Chromecast have further integrated streaming into the home entertainment experience.

The future of streaming looks toward advancements like virtual reality (VR) and augmented reality (AR), offering even more immersive entertainment experiences. Artificial intelligence (AI) and machine learning are also playing a role in content recommendation algorithms, enhancing user experience. Despite its success, the streaming industry faces challenges. Issues like copyright infringement, the digital divide, and concerns over data privacy and content censorship are ongoing discussions. Moreover, the increasing number of streaming services has led to 'subscription fatigue' among consumers.

The story of internet streaming is one of American ingenuity and its global impact. It has revolutionized how people consume media, democratized content creation, and reshaped the entertainment industry. As technology continues to evolve, the story of streaming remains an unfolding narrative of innovation, with American companies at its forefront, continuously influencing and shaping global media consumption.

NUCLEAR POWER

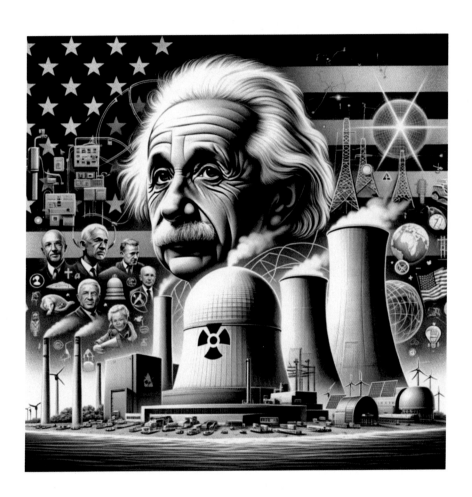

The Manhattan Project

Before nuclear power, the atom was primarily a subject of scientific curiosity. Pioneering work by scientists like Marie Curie and Albert Einstein laid the foundation for understanding the atom's potential for energy release. The story of nuclear power's development features American scientists, engineers, and entrepreneurs who recognized the possibilities of atomic energy. Innovators like Enrico Fermi, Robert Oppenheimer, and Hyman Rickover played pivotal roles in this journey. World War II accelerated nuclear research, leading to the creation of the Manhattan Project—a top-secret American initiative to develop nuclear weapons. The successful test of the atomic bomb in 1945 marked a turning point in the atomic age.

After World War II, the focus shifted from weapons to peaceful uses of nuclear energy. In 1951, the Experimental Breeder Reactor I (EBR-I) in Idaho became the first nuclear reactor to generate electricity, demonstrating the potential of nuclear power for civilian applications. In 1958, the Shippingport Atomic Power Station in Pennsylvania became the first full-scale nuclear power plant to provide electricity to the grid. This marked the beginning of commercial nuclear power generation in the United States. American engineers and scientists developed innovative reactor designs, such as the pressurized water reactor (PWR) and the boiling water reactor (BWR). These designs improved safety, efficiency, and reliability in nuclear power plants.

Nuclear power quickly gained popularity worldwide, offering a clean and efficient energy source. Countries like France and Japan embraced nuclear energy as a means of reducing dependence on fossil fuels and addressing growing energy demands. Nuclear energy brought both opportunities and challenges. Concerns about nuclear accidents, radioactive waste disposal, and nuclear proliferation led to debates and regulations aimed at ensuring safety and security.

Imagine a world without nuclear power. Energy production would rely more heavily on fossil fuels, exacerbating environmental issues like air pollution and climate change. Reducing carbon emissions and meeting energy demands would become even more challenging. Nuclear power played a crucial role in space exploration, providing reliable and long-lasting energy sources for spacecraft and deep space missions. Radioisotope thermoelectric generators (RTGs) powered missions like Voyager and the Mars rovers. Nuclear technology also contributed to advancements in medicine, enabling diagnostic imaging techniques like X-rays and positron emission tomography (PET) scans. Radioactive isotopes have been used in cancer treatment and medical research.

The United States has been a leader in nuclear research and innovation. National laboratories like Los Alamos and Oak Ridge continue to advance nuclear science, exploring fusion energy, advanced reactors, and nuclear waste solutions. As we look to the future, nuclear energy holds promise as a low-carbon energy source that can help address climate change. Advanced reactor designs and technologies aim to make nuclear power safer, more efficient, and more accessible. The United States plays a significant role in global nuclear governance through organizations like the International Atomic Energy Agency (IAEA). American leadership is vital in promoting nuclear non-proliferation and safe nuclear practices.

The development of nuclear power, driven by American innovation, has reshaped the world's energy landscape. It has provided clean, reliable energy and contributed to advancements in science, medicine, and space exploration. Nuclear energy represents the power of American ingenuity to unlock the potential of the atom for the benefit of humanity. As we celebrate the impact of nuclear power on our global society, we must also recognize the responsibility that comes with this technology. It is up to us to ensure that nuclear energy continues to be a safe, secure, and sustainable source of power, meeting the energy needs of the present and future generations.

PHONOGRAPH

The Silent Era

Imagine a world without recorded music, where the joy of melody and speech could only be experienced in the moment. This chapter explores the invention of the phonograph by Thomas Edison, highlighting its American origins and its profound impact on the music and entertainment industry worldwide. Before the phonograph, sound could only be heard through live performances, making it ephemeral and location-bound. The challenge was to capture and reproduce sound, and this quest led to the invention of the phonograph.

The story of the phonograph begins with Thomas Edison, an American inventor known for his relentless pursuit of innovation. Edison had a vision to bring recorded sound to the masses, revolutionizing the way people experience music and spoken word. In 1877, Edison unveiled the phonograph, a remarkable device capable of recording and reproducing sound. This groundbreaking invention marked a pivotal moment in the history of audio technology. The phonograph operated on a simple principle: a vibrating stylus engraved sound vibrations onto a rotating cylinder covered in tinfoil, and then played them back by reversing the process. It was a marvel of engineering.

Edison's invention had an immediate and profound impact on society. It transformed the way people enjoyed music, enabled the preservation of spoken word, and created new opportunities for entertainment. The phonograph paved the way for the modern music industry. Record labels, recording studios, and artists emerged to meet the growing demand for recorded music. It led to the popularization of genres and the rise of musical legends. Imagine a world without recorded sound. Music would be limited to live performances, and the spoken word would vanish into the air. The absence of recorded sound would alter the course of entertainment, culture, and communication.

The phonograph was just the beginning. Over the decades, audio recording technology evolved from tinfoil cylinders to vinyl records, magnetic tape, compact discs, and digital formats. Each advancement brought new possibilities and quality improvements. Thomas Edison's contributions extended beyond the phonograph. He played a pivotal role in the development of motion pictures, further influencing the entertainment industry and expanding the reach of American innovation. The invention of the phonograph exemplifies American innovation and its global impact. Edison's vision and ingenuity shaped the way the world experiences sound and entertainment. The phonograph not only changed the way people listened to music but also influenced culture. It became a symbol of progress, a source of nostalgia, and a catalyst for social change.

As technology advanced, sound recording transitioned from analog to digital formats. The digital revolution democratized music production, distribution, and consumption, bringing new challenges and opportunities to the industry. Thomas Edison's legacy serves as a reminder of the power of innovation to transform society. His relentless pursuit of progress and his willingness to challenge the status quo continue to inspire inventors and creators worldwide. As we look to the future, sound technology continues to evolve. Virtual reality, immersive audio, and artificial intelligence promise to redefine the way we experience sound and entertainment. The invention of the phonograph by Thomas Edison is a testament to American ingenuity and its profound impact on the music and entertainment industry. It brought music and spoken word to the masses, preserved cultural heritage, and transformed the way we experience sound. As we celebrate the impact of this invention on our global society, we must also recognize the enduring legacy of innovation and the endless possibilities it continues to unlock.

TESLA CAR

Electrifying the Future: The Invention of the Tesla Electric Car

In the realm of modern transportation, there emerged an American dream—one that aimed to revolutionize the way we move from point A to point B. This dream, which began with the creation of the Tesla electric car, would go on to reshape the automotive industry and leave a profound impact on the world.

At the forefront of this revolution stands Elon Musk, an American entrepreneur with a vision for a sustainable future. Musk, born in South Africa, moved to Canada, and educated in the United States, co-founded Tesla, Inc. in 2003 with a mission to accelerate the world's transition to sustainable energy. His ambitious goal: to create electric vehicles that would not only outperform traditional gasoline cars but also reduce the world's reliance on fossil fuels and combat climate change.

In 2008, Tesla introduced the world to the Tesla Roadster—the first production electric car to use lithium-ion battery cells. This sleek and sporty vehicle could accelerate from 0 to 60 mph in just 3.7 seconds, proving that electric cars could be both high-performance and environmentally friendly. The Tesla Roadster marked a turning point in the perception of electric vehicles. It was no longer just an eco-friendly choice; it was a choice for those who loved speed and innovation.

With the introduction of the Roadster, Tesla showed the world that electric cars could be more than just functional— they could be exciting. This sparked a wave of interest in electric vehicles, prompting other automakers to invest in electric car technology and launch their own electric models. In 2012, Tesla unveiled the Model S, a luxury electric sedan that combined cutting-edge technology with performance. It offered a range that surpassed any other electric car on the market, and its innovative Autopilot features foreshadowed the future of autonomous driving.

The Model S was not just a car; it was a statement of American innovation and a symbol of Tesla's commitment to pushing the boundaries of what electric vehicles could achieve. To support its growing demand, Tesla began building Gigafactories—massive manufacturing facilities dedicated to producing batteries, electric vehicles, and sustainable energy products. These Gigafactories, strategically located in the United States and around the world, represent a monumental investment in clean energy technology.

Tesla continued to expand its vehicle lineup with the introduction of the Model X, a stylish electric SUV with distinctive falcon-wing doors, the Model 3, a more affordable electric sedan aimed at a broader market, the Model Y, a compact SUV, and recently the CyberTruck, which is set to revolutionize the worldwide auto industry with its futuristic slick design and impressive capabilities. It signaled Tesla's ambition to disrupt not only the passenger car market but also the truck industry. These vehicles brought Tesla's electric technology to a wider audience. As Tesla's Model 3 gained popularity, it became one of the best-selling electric cars in the world. Its success demonstrated that electric cars could compete with traditional gasoline vehicles in price, range, and performance. Tesla's Supercharger network, a network of fast-charging stations, made long-distance travel in electric cars a reality.

Tesla's influence extended far beyond American borders. Its electric vehicles and clean energy products were sought after worldwide. Countries around the globe embraced Tesla's mission of sustainability, investing in electric vehicle infrastructure and offering incentives to promote electric car adoption. Tesla expanded its scope beyond vehicles, introducing the Powerwall—a home battery system that stores solar energy for use during peak hours or emergencies. This innovation allowed homeowners to reduce their reliance on the grid and harness clean energy from the sun.

One of Tesla's boldest promises was the development of fully autonomous driving technology. Through a series of software updates and over-the-air improvements, Tesla vehicles gained increasingly advanced Autopilot features, bringing the dream of self-driving cars closer to reality. Tesla's commitment to sustainability extended to its energy products. Solar panels, solar roofs, and utility-scale energy storage solutions were developed to help reduce our reliance on fossil fuels and transition to clean energy sources. Tesla's Powerpacks and Megapacks helped stabilize electrical grids and store renewable energy. It faced numerous challenges along its journey, from production bottlenecks to skepticism about electric vehicles. However, it persevered and achieved milestones that few thought possible, including becoming the world's most valuable automaker.

The story of Tesla is a testament to American innovation, ambition, and the pursuit of a sustainable future. It represents a new era in transportation—one where electric vehicles are at the forefront, and the automotive industry is redefined. As we look ahead, the future of transportation is electric, autonomous, and sustainable. Tesla's impact on the world goes beyond cars; it's about reshaping how we think about energy, transportation, and our responsibility to the planet. The invention of the Tesla electric car, led by American entrepreneur Elon Musk, has had a profound impact on the world. It has not only transformed the automotive industry but also accelerated the transition to sustainable energy and transportation.

From the groundbreaking Roadster to the Model S, Model 3, Model Y, Model X, and Cybertruck, Tesla has consistently pushed the boundaries of what electric vehicles can achieve. Its commitment to clean energy, battery technology, and autonomous driving has set the stage for a future where electric cars are the norm, and our planet's well-being is at the forefront of transportation. As we journey into the future, one thing is clear: the Tesla electric car is not just a mode of transportation; it's a symbol of innovation, sustainability, and the American spirit of progress.

SMART PHONE

iPhone Revolution

Before the smartphone era, mobile phones were seen as tools for basic communication. They served their primary purpose: making voice calls and sending text messages. However, they were far from the multifunctional devices we know today. The early mobile phones were large, heavy, and lacked the computing power to handle advanced tasks. The smartphone revolution began with the recognition that mobile phones could be much more than mere communication devices. American companies, driven by visionary leaders, saw the potential to create a new category of devices that could integrate various functions into a single, pocket-sized device.

Steve Jobs and Apple's iPhone

One of the defining moments in the history of smartphones was the introduction of the iPhone by Apple in 2007. Steve Jobs, the co-founder of Apple, unveiled a device that would change the way we interact with technology forever. The iPhone combined a mobile phone, an iPod for music, and an internet communicator, all in one sleek and intuitive package. The iPhone's impact was immediate and profound. Its design was groundbreaking, featuring a large touchscreen that replaced physical buttons and a user interface that was easy to navigate. It marked a departure from the clunky, button-filled phones of the past and set a new standard for aesthetics and functionality. One of the iPhone's most significant innovations was the introduction of the App Store in 2008. This digital marketplace allowed developers to create and distribute applications (apps) for the iPhone. It opened the floodgates to creativity and innovation, transforming smartphones into versatile tools that could cater to a wide range of user needs.

The App Store quickly became a hub for developers, offering a platform to reach millions of users. This ecosystem of apps expanded the capabilities of smartphones exponentially. Users could now do more than just make calls and send messages; they could play games, access news, edit photos, manage finances, and much more, all from their smartphones.

The Global Impact

The iPhone's success had a ripple effect around the world. It triggered a global smartphone boom, with other manufacturers and operating systems entering the market. Companies like Google introduced the Android operating system, providing an alternative to Apple's iOS. This competition drove innovation, leading to the creation of diverse devices and features. Smartphones have become a ubiquitous part of our daily lives. People of all ages, backgrounds, and professions rely on them for communication, entertainment, productivity, and information. They have transformed industries such as photography, music, and gaming. The ability to capture high-quality photos and videos, listen to music on the go, and play immersive games has become a standard feature of modern smartphones.

Smartphones have also revolutionized the way we access information. The internet is now readily available at our fingertips, allowing us to browse websites, check emails, and access social media from virtually anywhere. This easy access to information has reshaped how we consume news, stay connected with friends and family, and participate in online communities.

Imagine a world without smartphones. Communication would revert to being less immediate, with people relying on landlines or computers to connect. Access to information would be limited to specific locations with internet access, making real-time updates and communication less accessible. The absence of smartphones would have a significant impact on various industries. E-commerce, for example, relies heavily on mobile apps and websites to reach customers and facilitate online shopping. The convenience of ordering products and services from your smartphone has transformed the way we shop and conduct business. Entertainment, too, would feel the void left by smartphones. Streaming services, mobile games, and on-the-go entertainment have become an integral part of our leisure time. Smartphones have enabled us to watch movies, play games, and listen to music wherever and whenever we choose. Smartphones have also become valuable tools in education. They offer access to vast amounts of information and interactive learning experiences. Educational apps, online courses, and digital textbooks have made learning more accessible and engaging. Smartphones have the potential to bridge educational gaps and provide opportunities for lifelong learning.

In healthcare, smartphones have played a crucial role in enabling telemedicine. They allow patients to consult with healthcare professionals remotely, providing access to medical care even in remote areas. Health tracking apps and wearables have empowered individuals to monitor their health and well-being, leading to more proactive healthcare management.

The Future of Smartphones

As we reflect on the impact of smartphones, we must also look ahead to the future. The evolution of these devices shows no signs of slowing down. Advancements in 5G technology promise faster and more reliable connectivity, enabling new possibilities in areas like augmented reality and virtual reality. Artificial intelligence (AI) is another frontier that smartphones are actively exploring. AI-powered features, such as voice assistants and predictive algorithms, are already enhancing the user experience. As AI continues to advance, smartphones may become even more intuitive and capable of understanding and anticipating user needs.

Virtual reality (VR) and augmented reality (AR) are poised to transform how we interact with the digital world. VR immerses users in virtual environments, while AR overlays digital information onto the real world. These technologies have applications beyond entertainment and gaming. They can enhance training and education, improve remote collaboration, and revolutionize industries like architecture and healthcare. As smartphones continue to evolve and integrate into every aspect of our lives, it's essential to consider the ethical and societal implications. Privacy, data security, and the impact of technology addiction are topics that deserve careful consideration. Ensuring that smartphones are used for the betterment of humanity while respecting individual rights and well-being is a responsibility that falls on both users and technology companies.

The smartphone, with its American origins, stands as a symbol of innovation and technological leadership. It represents the United States' commitment to pushing the boundaries of what is possible, even in the face of complex challenges. American companies have played a central role in shaping the digital landscape and connecting people worldwide. The impact of the smartphone extends beyond individual devices. It has fueled the growth of tech ecosystems, created jobs, and driven economic growth. American technology companies have become global giants, influencing not only how we communicate but also how we work, play, and interact with the world.

The invention of the smartphone has fundamentally transformed the world. It has redefined how we communicate, work, and interact with the world around us. The smartphone represents the power of technology to shape the future and connect humanity on a global scale. As we celebrate the impact of the smartphone on our global society, we must also recognize the responsibility that comes with this technology. It has the power to unite, inform, and empower, and it is up to us to ensure that it is used for the betterment of all humanity. In a world where connectivity is more critical than ever, the smartphone remains a powerful tool for positive change and progress.

INTERNET SEARCH ENGINE

New Verb – Google It

The Democratization of Knowledge

Before the advent of search engines, the internet was a chaotic realm of information. Finding specific content was akin to searching for a needle in a haystack. The challenge was to create a tool that could organize and index the ever-growing digital landscape. The story of internet search engines begins with American innovators who recognized the need for a better way to navigate the web. Pioneers like Larry Page, Sergey Brin, and others played pivotal roles in shaping the search engine landscape.

In 1998, Larry Page and Sergey Brin founded Google, a company that aimed to revolutionize how people find information on the internet. Their search engine's algorithm, PageRank, prioritized relevant search results, setting a new standard for accuracy and efficiency. Google's success led to the development of a suite of innovative products, including Gmail, Google Maps, and Google Drive. The company's mission was to organize the world's information and make it universally accessible and useful.

Google quickly became the dominant search engine globally. Its impact extended beyond search, influencing advertising, online business models, and digital marketing strategies. Imagine a world without internet search engines. Navigating the web would be cumbersome, and accessing information would be a slow and frustrating process. The absence of search engines would hinder progress in education, research, and business. Internet search engines democratized access to knowledge. They made information available to anyone with an internet connection, breaking down barriers to education and creating opportunities for self-improvement.

The invention of internet search engines exemplifies American innovation and its global impact. American tech companies, like Google, have become synonymous with the digital age, shaping the way societies access and share information. Search engines have faced challenges related to privacy, data security, and the spread of misinformation. American tech companies have worked to address these issues while upholding principles of free expression and open access to information. As we look to the future, internet search engines will continue to evolve. Advancements in artificial intelligence, voice search, and personalized recommendations promise to enhance the search experience and make information even more accessible.

The success of American internet search engines has contributed to the growth of tech ecosystems, creating jobs, driving economic growth, and fostering innovation across various industries. The invention of internet search engines, driven by American innovation, has connected people across the globe. It has transformed how we learn, work, and interact with the vast digital landscape. The invention of internet search engines, exemplified by companies like Google, is a testament to American ingenuity and its profound impact on the world. It has revolutionized the way we access information, conduct research, and navigate the digital realm. As we celebrate the impact of this invention on our global society, we must also recognize the responsibility that comes with this technology. It is up to us to ensure that search engines continue to be a tool for knowledge, discovery, and progress while addressing the challenges of the digital age.

SOCIAL MEDIA

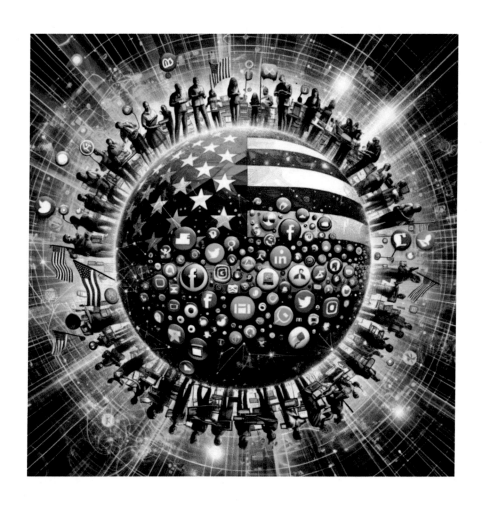

In the 21st century, a digital revolution began that would change the way people connect, communicate, and share their lives. Before the rise of social media, the internet was primarily a place for information and limited interaction. Online communities and forums existed, but they lacked the real-time, user-friendly, and visually engaging platforms that would soon become synonymous with social media.

The story of social media begins with American innovators who recognized the potential of the internet to facilitate social interaction. Visionaries like Mark Zuckerberg, Jack Dorsey, and others played pivotal roles in shaping the landscape of digital socialization. In 2004, Mark Zuckerberg and his college roommates launched Facebook, a platform initially designed for college students to connect with each other. Its user-friendly interface and focus on real identities set it apart and laid the foundation for a global phenomenon.

Facebook's success paved the way for the emergence of various social networks, each with its own unique features and purposes. Platforms like Twitter(X), LinkedIn, and Instagram quickly gained popularity, offering users diverse ways to connect and share. Social media platforms transcended borders, languages, and cultures, becoming a global phenomenon. People worldwide embraced social media as a means of staying connected with friends and family, sharing their thoughts and experiences, and engaging with a broader community. As social media platforms flourished, they faced challenges related to privacy, content moderation, and the spread of misinformation. American technology companies navigated these complex issues while adhering to principles of free expression and innovation.

Social media fundamentally transformed the way people

communicate. It introduced real-time, global communication, enabling users to connect with others instantly, regardless of geographical boundaries. It reshaped the nature of relationships and interpersonal interactions. The birth of social media marked a pivotal moment in the history of human communication. It unleashed a digital revolution that redefined how we connect with others, share our lives, and interact with the world.

The story of social media is intrinsically tied to American innovation. From the birth of platforms like Facebook and Twitter to the development of advanced algorithms and technologies, American companies and entrepreneurs played a central role in shaping the digital social landscape. This commitment to innovation and the pursuit of groundbreaking ideas helped the United States maintain a leadership position in the tech industry.

Social media's impact extended far beyond the United States. It became a global phenomenon, connecting people from diverse cultures, languages, and backgrounds. Through platforms like Facebook, Twitter(X), and Instagram, individuals could interact with others from around the world, fostering cross-cultural understanding and promoting cultural exchange. Social media empowered individuals to have a voice on a global stage. It democratized the ability to share opinions, advocate for causes, and hold institutions accountable. Furthermore, social media provided a platform for underrepresented voices and marginalized communities. It allowed individuals to share their stories, experiences, and perspectives, challenging traditional narratives and amplifying voices that had previously been unheard. This inclusivity contributed to a more diverse and vibrant digital landscape.

As social media's influence grew, so did the responsibility of technology companies to address its challenges. Companies like Facebook, Twitter(X), and Google faced scrutiny over issues of privacy, content moderation, and the spread of misinformation. They navigated complex ethical and operational dilemmas while striving to uphold principles of free expression and innovation. Efforts to combat misinformation and enhance content moderation became central to the social media landscape. Fact-checking initiatives, algorithmic adjustments, and partnerships with external organizations aimed to mitigate the impact of false or harmful content. Social media companies also implemented features to promote digital well-being, such as time management tools and anti-bullying measures.

The story of social media is one of continuous evolution and adaptation. It reflects the ever-changing nature of technology and its profound impact on society. From its American origins to its global reach, social media has reshaped how we connect, communicate, and share in the digital age. While technology companies bear responsibility for the platforms they create and maintain, users also play a crucial role in shaping the social media landscape. Responsible and ethical use of social media is essential for maintaining a healthy digital ecosystem.

The invention of social media represents American innovation and its role in shaping the digital landscape. From its early days as a platform for college students to its global reach and influence, social media has transformed how societies communicate and connect. Social media's impact is felt in areas ranging from politics and activism to business and personal relationships. It has empowered individuals to have a voice, mobilized movements, and fostered cultural exchange.

However, it has also posed challenges related to privacy, misinformation, and digital well-being.

As we navigate the ever-evolving landscape of social media, it is essential to recognize its potential for positive change while addressing its complexities. Technology can be a powerful force for connectivity, expression, and progress, but it requires thoughtful consideration, responsible use, and ongoing innovation to harness its full potential for the betterment of society. The story of social media serves as a reminder that, in the digital age, our choices and actions online have a profound impact on the world we share.

DIGITAL CAMERA

Capturing the Future

Imagine a world without digital cameras, where capturing a moment required film, darkrooms, and patience. Before digital cameras, photography relied on chemical processes and physical film. Each photo was a tangible print, and the process was time-consuming and resource-intensive. The challenge was to create a device that could capture, store, and share images electronically. The story of the digital camera begins with American innovators who recognized the potential of digital technology to transform photography. Pioneers like Steve Sasson and Eugene Lally played pivotal roles in shaping the digital camera landscape.

In 1975, engineer Steve Sasson, working at Eastman Kodak, created the world's first digital camera. It was a bulky, experimental device that used magnetic tape to record black-and-white images, but it laid the foundation for a revolution in photography. The digital camera evolved rapidly, with companies like Kodak, Nikon, and Canon leading the way. Innovations in sensor technology, image processing, and storage transformed digital cameras into versatile and accessible tools.

Digital cameras democratized photography. They made it easy for anyone to take, edit, and share high-quality photos, breaking down barriers to entry and creating a new era of visual storytelling. The integration of cameras into smartphones, led by American tech giants like Apple and Google, further revolutionized photography. It made capturing and sharing moments an integral part of daily life. Imagine a world without digital cameras. Photography would remain a niche hobby, and the instant sharing of images would be impossible. The absence of digital cameras would impact journalism, art, and personal expression. Digital cameras also transformed professional photography. They offered greater flexibility, efficiency, and quality, allowing photographers to experiment, innovate, and push the boundaries of their craft. The invention of digital cameras exemplifies American innovation and its global impact. American tech companies, like Kodak, played pivotal roles in shaping the digital imaging industry and influencing societies worldwide. Digital photography brought challenges related to privacy, copyright, and image manipulation. American companies and legal frameworks have grappled with these issues while preserving creative expression and innovation.

As we look to the future, digital imaging will continue to evolve. Advancements in computational photography, virtual reality, and augmented reality promise to redefine how we capture and experience visual content. The invention of the digital camera, driven by American innovation, has connected people across the globe through the universal language of images. It has transformed how we document our lives, share our stories, and express our creativity. The invention of the digital camera is a testament to American ingenuity and its profound impact on the world of photography and visual communication. It has democratized the art of capturing moments and has enriched our lives with the power of visual storytelling. As we celebrate the impact of this invention on our global society, we must also recognize the responsibility that comes with this technology. It is up to us to ensure that digital imaging continues to be a tool for creativity, communication, and preservation while addressing the challenges of the digital age.

CREDIT CARD

The invention of the credit card is a story deeply rooted in American innovation and entrepreneurship, a tale that spans several decades and reflects the changing dynamics of the American economy and society. This story, unfolding in the heart of the 20th century, not only revolutionized how Americans conducted transactions but also had a profound impact on the global economy. Post World War II America was a period of unprecedented economic growth. The country emerged from the war as a global superpower with a booming economy. People were moving to the suburbs, buying cars, and living a lifestyle that required more flexible financial systems. It was in this backdrop that the first seeds of the credit card idea were planted.

The story of the credit card often begins with a man named Frank McNamara. McNamara, a New York businessman, found himself in an embarrassing situation one evening in 1949. After dining at a New York City restaurant, he realized he had left his wallet in another suit. This incident led him to conceive an idea: a card that would allow customers to charge the cost of meals at various establishments. In 1950, along with his partner Ralph Schneider, McNamara founded Diners Club. Their first 'credit card' was a cardboard card, which 200 select members could use in 27 restaurants in New York. The concept was simple yet revolutionary: Members could dine without cash, and the establishments were guaranteed payment by Diners Club, which in turn, would settle the bill with the members later.

The success of Diners Club was the beacon for other companies. American Express, then primarily a traveler's cheque company, jumped into the fray. In 1958, under the leadership of Ralph Reed, American Express launched its own credit card. This move was significant – it marked the transition of credit cards from a dining convenience to a broader financial tool used for various purchases. The American Express card, made of plastic (a novelty then), quickly gained popularity. By the end of 1958, it had been used in thousands of transactions, both in the United States and abroad, signaling the first international use of credit cards.

While Diners Club and American Express were making strides, another revolution was brewing in the banking sector. In 1958, Bank of America released the BankAmericard in Fresno, California. This was the first bank-issued credit card intended for general consumer use, not just for travel and entertainment. The BankAmericard system marked a significant innovation: the revolving credit model. This allowed consumers to carry a balance from month to month, paying interest on the unpaid amount. The concept rapidly spread across the country, leading to the formation of Visa in the 1970s.

Another pivotal development was the creation of the Interbank Card Association (now MasterCard) in 1966. This was a group of banks that banded together to accept each other's cards, creating a larger network and giving birth to the modern credit card industry. The evolution of credit cards was heavily influenced by technological advancements. Magnetic stripes, electronic authorization, and computerized billing were just a few of the innovations that streamlined credit card use and made it more secure. However, with the rise of credit cards, there was also a need for regulation. The U.S. government stepped in with laws like the Truth in Lending Act (1968) and the Fair Credit Billing Act (1974), protecting consumers from unfair billing practices and establishing the concept of consumer rights in financial services.

The concept of the credit card, born in America, quickly spread worldwide. Countries across Europe, Asia, and beyond started adopting the credit card model, tailoring it to their own economic and cultural contexts. The credit card became a symbol of financial modernity and convenience, breaking down traditional barriers in global commerce. The invention and evolution of the credit card are quintessentially American stories. They reflect the country's spirit of innovation, its dynamic economy, and its role as a global trendsetter. The credit card transformed how people access and spend money, not just in America but around the world. It facilitated a move towards a cashless society, enabled the rise of e-commerce, and forever changed consumer habits and financial services.

As the 20th century gave way to the 21st, the credit card industry faced a new frontier: the digital revolution. The advent of the internet and e-commerce demanded further innovation in the credit card realm. The industry responded with advancements like online account management, virtual credit card numbers for online shopping, and enhanced security features like chip technology and contactless payments. Security concerns, particularly with the rise of online transactions, led to significant technological advancements. The introduction of EMV chip technology, pioneered in Europe and later adopted in the U.S., greatly reduced fraud and counterfeiting of cards. This shift also paved the way for contactless payments, using Near Field Communication (NFC) technology, allowing for faster and more secure transactions.

The proliferation of smartphones brought another evolution in credit card technology. Mobile payment platforms like Apple Pay and Google Wallet transformed smartphones into digital wallets, allowing users to store credit card information securely on their phones and use them for contactless payments. This integration of technology continued to blur the lines between traditional banking and the digital world. As credit cards became more embedded in daily life, regulatory challenges emerged. Issues like identity theft, data breaches, and privacy concerns became paramount. In response, regulations and industry standards evolved, aiming to protect consumers and ensure the integrity of the financial system. The Payment Card Industry Data Security Standard (PCI DSS) was established, setting security standards for all entities that handle credit card data.

The globalization of the credit card industry saw American financial institutions expanding their reach worldwide. However, this expansion wasn't without challenges. Different countries had varied financial practices, cultural attitudes towards credit, and regulatory environments. American companies had to adapt, tailoring their services to local markets while maintaining global standards. The widespread use of credit cards had a significant impact on consumer behavior and the global economy. It facilitated a shift from saving to spending, fueling consumerism. Credit cards also played a critical role in financial inclusion, providing many people with access to credit for the first time, which in turn spurred economic growth. The credit card also brought about societal changes. It altered the psychology of spending, making transactions more abstract and, in some cases, leading to increased consumer debt. This necessitated financial education and awareness as integral components of consumer finance.

The future of credit cards is intertwined with the broader fintech revolution. Technologies like blockchain, artificial intelligence, and biometric authentication are set to redefine the security and functionality of credit cards. The industry is poised for further innovation, with a focus on enhanced security, user experience, and integration with emerging digital ecosystems. The story of the credit card is one of continual evolution, reflecting broader trends in technology, society, and the global economy. It's a story that underscores the power of American innovation and its global impact. From a simple piece of cardboard in the hands of a few diners in New York to a sophisticated digital tool used by billions worldwide, the credit card remains a symbol of financial innovation and a facilitator of global commerce. As we stand on the cusp of new advancements in financial technology, the credit card's journey from a novel idea to a global financial instrument is a testament to the enduring spirit of innovation and adaptability that continues to drive economic and societal change.

ATM

Banking Beyond Borders

Imagine a world without Automated Teller Machines (ATMs), where accessing cash or conducting banking transactions required visiting a bank during its operating hours. Before the ATM, banking was often a time-consuming process that required trips to the bank, long queues, and limited access to funds. The challenge was to create a device that could provide 24/7 banking convenience.

The story of the ATM begins with American innovators who recognized the potential of automated banking services. Pioneers like Don Wetzel, John Shepherd-Barron, and George Simjian played pivotal roles in shaping the ATM landscape. In the late 1960s and early 1970s, engineers like Don Wetzel and his team at Docutel Corporation developed the world's first successful ATM prototypes. These machines allowed users to withdraw cash, check balances, and perform basic banking transactions. American banks quickly adopted ATM technology, and the machines began to appear in urban centers. The idea of 24/7 access to banking services resonated with consumers and led to widespread adoption.

Over time, American companies like Diebold and NCR worked to miniaturize ATMs and make them accessible to a broader range of financial institutions and consumers. Imagine a world without ATMs. Accessing cash and conducting banking transactions would be limited to banking hours, impacting financial accessibility and convenience for individuals and businesses. The invention of the ATM transformed global banking. It allowed for greater financial inclusion, enabling people to access banking services in remote areas and during non-business hours. The invention of the ATM exemplifies American technological innovation and its global impact. American companies have played pivotal roles in shaping the ATM industry and influencing banking practices worldwide. ATMs brought challenges related to security, fraud prevention, and user privacy. American financial institutions and regulators have worked to address these issues while maintaining the benefits of banking automation.

As we look to the future, banking technology continues to evolve. Advancements in biometrics, mobile banking, and digital payments promise to redefine how we conduct financial transactions. The success of the ATM has contributed to a transformed financial landscape, fostering financial accessibility, convenience, and efficiency. The invention of the ATM, driven by American innovation, has connected people across the globe through the shared experience of convenient and accessible banking. It has transformed financial institutions, customer expectations, and the way we manage our finances.

The invention of the ATM is a testament to American technological ingenuity and its profound impact on the world of banking and financial accessibility. It has democratized banking services, making them available to a broader range of people and businesses. As we celebrate the impact of this invention on our global society, we must also recognize the responsibility that comes with this technology. It is up to us to ensure that ATMs continue to be a tool for financial inclusion, convenience, and security while addressing the challenges of modern banking.

TRAFFIC LIGHT

The story of the traffic light begins in the late 19th and early 20th centuries, a time when the world was experiencing a significant shift in transportation. With the rise of automobiles, the streets that were once dominated by horses and carriages were now shared with motor vehicles. This transition created a need for better traffic management systems, especially in bustling American cities.

Before the advent of the electric traffic light, police officers would often control traffic at busy intersections. However, this method was not only labor-intensive but also ineffective in managing the increasing number of vehicles. The first known traffic signal was invented in 1868 by J.P. Knight, a British railway signaling engineer. It was a gas-lit signal installed outside the Houses of Parliament in London. Unfortunately, this early attempt resulted in an explosion and was deemed impractical for broader use.

The real breakthrough came from the United States. The first electric traffic signal was developed by James Hoge in 1914 and installed in Cleveland, Ohio. Hoge's system used red and green lights and was wired to a manually operated switch located inside a control booth. This invention allowed police and fire stations to control the flow of traffic in response to emergencies, which was a significant advancement in urban traffic control. Another key figure in the development of traffic lights was Garrett Morgan, an African American inventor and businessman. In 1923, Morgan patented a traffic control device, which was a T-shaped pole unit that featured three positions: Stop, Go, and an all-directional stop position. This third position halted traffic in all directions to allow pedestrians to cross streets safely. Morgan's design was a precursor to the modern three-way traffic signal.

As automobiles became more prevalent, the need for a standardized traffic control system became apparent. The

American Association of State Highway Officials played a crucial role in this process, leading to the standardization of traffic lights as we know them today - with red, yellow, and green lights. The next significant development in traffic light technology was the automation of traffic signals. The first automatic traffic lights were installed in 1922 in Houston, Texas. These lights used timers to switch between signals, which greatly improved traffic flow and reduced the need for manual operation. The success of traffic lights in the United States led to their rapid adoption around the world. Countries across the globe began installing traffic lights to manage the increasing flow of vehicles and improve road safety. The American model of the three-color traffic light became the standard internationally.

Over the years, technological advancements have continued to enhance traffic light systems. The introduction of computer programming and sensor technology allowed for the development of smart traffic lights, which could adapt to changing traffic conditions in real time. These advancements, many pioneered in the United States, have made traffic management more efficient and responsive. With the growing awareness of environmental issues, the focus shifted towards making traffic lights more energy-efficient. The adoption of LED lights, which consume less power and have a longer lifespan than traditional bulbs, was a significant step in this direction. This change not only reduced energy consumption but also lowered maintenance costs.

Looking towards the future, traffic light technology is set to become even more sophisticated. Innovations such as integration with vehicle-to-infrastructure communication and AI-based traffic management systems are on the horizon. These advancements promise to further enhance traffic efficiency and safety, potentially reducing congestion and emissions.

The invention and evolution of the traffic light are emblematic of American ingenuity and its ability to address complex challenges. From Hoge's and Morgan's pioneering efforts to the high-tech traffic management solutions of today, the traffic light's journey reflects America's spirit of innovation and problem-solving. As a tool that has become indispensable in cities worldwide, the traffic light stands as a testament to an American invention that has had a profound and lasting impact on global society.

ZIPPER

Unzipping Innovation: The American Origins of the Zipper

In the early 20th century, a small yet significant invention changed the way people fastened their clothing and belongings. This chapter explores the invention of the zipper, a practical and versatile device that made its mark on the world, thanks to American ingenuity.

The zipper's story begins with Whitcomb L. Judson, an American inventor who was frustrated with the time-consuming process of fastening shoes and clothing. In 1891, he patented the "clasp locker," an early version of the zipper, with the goal of creating a faster and more efficient fastener. Judson's invention laid the groundwork for future developments. In 1913, Swedish-American engineer Gideon Sundback improved upon Judson's design, creating a more reliable and efficient fastener that resembled the modern zipper. This marked a significant step forward in zipper technology.

The zipper's potential was recognized by the American manufacturer Talon, which began producing zippers for various applications. World War I accelerated the zipper's adoption, as it was used in military gear, leading to increased awareness and acceptance among the general public. Zippers made their way into various industries, from the fashion world to aviation and sportswear. Fashion designers embraced zippers as a stylish and functional addition to clothing. They provided a new level of convenience and design possibilities.

The distinctive sound of a zipper has become synonymous with fastening and unfastening clothing. It's a sound that evokes memories and nostalgia, reminding us of the practicality and efficiency that zippers bring to our lives. Zippers have become an integral part of our daily routines.

From jackets and pants to backpacks and luggage, zippers keep our belongings secure and easily accessible. They simplify our lives in ways we often take for granted. The invention of the zipper represents American innovation at its finest. It transformed the way people fasten clothing and accessories, and its impact has been felt worldwide. The zipper is a symbol of American ingenuity that has benefited people across the globe.

Imagine a world without zippers. Clothing would be harder to put on and take off. Luggage and bags would lack secure closures. The efficiency and convenience that zippers bring to our lives would be sorely missed. The zipper's legacy lives on as a testament to American inventiveness. It continues to evolve, with advancements in materials and design. Zippers are now found in an array of applications, from high-performance outdoor gear to space exploration suits. Fashion designers have incorporated zippers into their creations in innovative ways. Zippers have become not just functional but also fashionable, serving as decorative elements and design statements. They add a touch of modernity and edginess to clothing and accessories. The zipper's impact extends to the automotive and aviation industries, where it is used in vehicle seating and interior design. Zippers provide access to hidden compartments and facilitate maintenance and repair.

As we look to the future, zippers will continue to play a vital role in our lives. Advancements in materials, such as water-resistant and self-healing zippers, promise new levels of functionality. The zipper's journey of innovation is far from over. Zippers have become symbols of individuality and self-expression. They can be used to personalize clothing and accessories, allowing people to showcase their unique style and personality. Zippers have also found their way into the world of technology, where they are used in everything from smartphone cases to wearable tech. Zippers provide easy access to devices and protect them from the elements.

The zipper is an American invention that has left an indelible mark on the world. It has simplified our lives, revolutionized fashion, and become a symbol of efficiency and convenience. The next time you hear that familiar zipper sound, remember the American innovators who brought this everyday marvel into our lives. With the zipper, American inventors transformed how we fasten our clothing and belongings, proving that even the simplest of innovations can have a profound and lasting impact on our daily lives.

TOILET PAPER

The Rise of Rolled Toilet Paper

In the realm of hygiene and comfort, one American invention stands out as both practical and ubiquitous: the roll of toilet paper. This chapter explores the fascinating story behind the creation of toilet paper, its American origins, and its profound impact on people's lives worldwide. Before the invention of toilet paper, people relied on various methods to cleanse themselves after using the toilet. Common practices included using leaves, rags, corn cobs, or even their bare hands. The need for a more sanitary and convenient solution was evident.

The story of toilet paper's invention begins with Joseph Gayetty, an American entrepreneur from New York. In 1857, he introduced "Gayetty's Medicated Paper," which was marketed as a pre-moistened sheet for personal hygiene. While not quite toilet paper as we know it today, it was a step in the right direction. The true evolution of toilet paper came with the contributions of Seth Wheeler, another American inventor. In 1871, Wheeler patented the concept of rolled and perforated toilet paper. His invention was a significant leap forward, making toilet paper more practical and easier to use. In 1890, the Scott Paper Company, founded by the Scott brothers—Clarence and E. Irvin—began mass-producing rolled and perforated toilet paper. Their product, known as "Sani-Tissue," was a game-changer in the world of hygiene and comfort.

The Scott brothers' invention quickly gained popularity, especially among the growing urban population. It brought a level of convenience and sanitation that people had never experienced before. Toilet paper became a household essential. Toilet paper's widespread adoption was a reflection of societal progress and the recognition of the importance of sanitation and hygiene. It became a symbol of modern living, representing a cleaner and more comfortable way of life. The invention of toilet paper is a testament to American innovation and ingenuity. It exemplifies the country's ability to address everyday challenges and improve people's lives through practical inventions. Without American inventors like Seth Wheeler and the Scott brothers, the world would be a very different place.

Imagine a world without toilet paper. People would still be relying on less sanitary methods, compromising hygiene and comfort. The absence of toilet paper would impact not only personal cleanliness but also public health. Toilet paper's impact extends far beyond American borders. It has become a universal necessity, embraced by people from diverse cultures and backgrounds. It transcends language barriers, providing a common thread of comfort and convenience. Toilet paper has continued to evolve over the years. It is now available in various types, including recycled, bamboo, and ultra-soft varieties. Innovations like flushable wipes and bidet attachments have further improved personal hygiene. As the world becomes more environmentally conscious, there is a growing awareness of the ecological impact of toilet paper production. Companies are exploring eco-friendly alternatives and sustainable sourcing to reduce their environmental footprint.

Toilet paper has even found its way into popular culture, humor, and art. It has been the subject of jokes, cartoons, and advertising campaigns. Its presence in our lives is so pervasive that we often take it for granted. As we look to the future, toilet paper will likely continue to evolve. Innovations in materials and production methods may lead to even more sustainable and environmentally friendly options. The quest for comfort and sanitation remains a timeless pursuit. Toilet paper's impact on our daily lives cannot be overstated. It is a small yet essential invention that has brought comfort and cleanliness to millions. Its story serves as a reminder of the power of innovation to improve the most fundamental aspects of human existence.

The invention of toilet paper, pioneered by American inventors like Seth Wheeler and the Scott brothers, has left an indelible mark on the world. It has elevated personal hygiene, brought comfort to countless lives, and become a symbol of progress and modernity. Toilet paper is a testament to the idea that even the simplest innovations can have a profound and lasting impact on our daily lives. Its American origins and global adoption highlight the power of practical inventions to improve the quality of life for people everywhere.

MICROWAVE OVEN

Cooking at the Speed of Light

Imagine a world without microwave ovens, where reheating leftovers, defrosting food, and cooking convenience meals took much longer. Before the microwave oven, cooking and heating food involved conventional ovens, stovetops, and time-consuming methods. The challenge was to create a device that could rapidly cook and heat food using microwave radiation.

The story of the microwave oven begins with American innovators who recognized the potential of microwave technology for culinary applications. Pioneers like Percy Spencer and Raytheon Corporation played pivotal roles in shaping the microwave oven landscape. In 1945, engineer Percy Spencer, working at Raytheon Corporation, made a serendipitous discovery. While testing a magnetron, he noticed that a candy bar in his pocket had melted. This observation led to the development of the first microwave oven. In 1947, Raytheon introduced the Radarange, the first commercially available microwave oven. It was a massive, expensive appliance primarily used in industrial and commercial settings.

Over the decades, American companies like Litton and Amana Corporation worked to miniaturize and make microwave ovens affordable for residential use. This accessibility transformed the way people cooked and reheated food at home. Imagine a world without microwave ovens. Food preparation would be more time-consuming, and the convenience of reheating leftovers would be lost. The absence of microwave ovens would impact kitchens, restaurants, and food industries. The microwave oven revolutionized culinary culture. It introduced new cooking techniques, convenience foods, and microwave-safe packaging. It also influenced food habits and the popularity of ready-to-eat meals.

The invention of the microwave oven exemplifies American innovation and its global impact. American tech companies have played pivotal roles in shaping the microwave oven industry and influencing culinary practices worldwide. Microwave cooking brought challenges related to food safety, nutrition, and taste. American researchers and regulators have worked to address these issues while promoting safe and healthy microwave cooking.

As we look to the future, microwave technology continues to evolve. Advancements in smart microwave ovens, precision cooking, and food science promise to redefine how we use this appliance in our kitchens. The success of the microwave oven has contributed to a kitchen revolution, simplifying cooking and changing how we approach meal preparation and convenience foods. The invention of the microwave oven, driven by American innovation, has connected people across the globe through the shared experience of faster and more convenient cooking. It has transformed kitchens, food industries, and culinary habits. The invention of the microwave oven is a testament to American ingenuity and its profound impact on the world of food preparation and culinary culture. It has brought convenience and speed to our kitchens, changing the way we cook and enjoy meals. As we celebrate the impact of this invention on our global society, we must also recognize the responsibility that comes with this technology. It is up to us to ensure that microwave ovens continue to be a tool for efficiency, convenience, and culinary creativity while addressing the challenges of modern cooking.

REFRIGERATOR

The quest for refrigeration began long before the modern refrigerator came into existence. Ancient civilizations used various methods to preserve food, but it was not until the 18th and 19th centuries that the foundations for modern refrigeration technology were laid. Early attempts involved using ice and salt mixtures to create lower temperatures, but these methods were rudimentary and not very efficient.

The real breakthrough in refrigeration technology came in the mid-19th century. An American physician, John Gorrie, played a pivotal role in this development. In the 1840s, Gorrie built a machine that used compression to make ice, aiming to cool the air for his yellow fever patients. Although Gorrie's machine was a prototype and not commercially successful, it laid the groundwork for future developments in refrigeration technology. The late 19th and early 20th centuries saw significant advancements in refrigeration technology, with several American inventors and engineers leading the way. Systems that used various gases for cooling began to emerge, and the commercial refrigeration industry started to take shape. One of the key figures in this era was an American engineer, Alexander Twining. In 1856, Twining received a patent for an ice-making machine that marked one of the earliest forms of mechanical refrigeration in the U.S.

The invention of the domestic refrigerator was a game-changer, bringing the technology into American homes. In 1913, Fred W. Wolf of Fort Wayne, Indiana, introduced one of the first domestic refrigerators, the "Domelre" (Domestic Electric Refrigerator). This early version was a crude unit that sat on top of an ice box, but it paved the way for future innovations. In the 1920s, the modern electric refrigerator began to take shape. General Electric, an American company, played a significant role in this development. In 1927, GE released the "Monitor-Top" refrigerator, which became immensely popular and made household refrigeration widely accessible in America.

The refrigerator had a profound impact on American lifestyle and culture. It changed the way people shopped, cooked, and ate. The ability to store food safely for longer periods led to less frequent shopping trips, a wider variety of available foods, and the emergence of frozen foods. The refrigerator became a symbol of modern convenience and efficiency in the American household. The influence of the American refrigerator soon spread globally. Countries around the world began adopting this essential appliance, transforming food storage and preservation methods. The refrigerator's impact on food safety, nutrition, and culinary habits was significant, contributing to improved health and lifestyle globally.

Over the years, American companies and engineers continued to innovate and improve refrigerator technology. Developments included more efficient cooling systems, the introduction of freon as a safer refrigerant, and the eventual phasing out of freon due to environmental concerns. In the late 20th and early 21st centuries, the focus shifted towards making refrigerators more energy-efficient and environmentally friendly. American companies were at the forefront of developing energy-saving models and using alternative, eco-friendly refrigerants.

The dawn of the digital age brought new innovations to refrigerator technology. Smart refrigerators, equipped with touch screens, internet connectivity, and internal cameras, emerged as the latest trend. These high-tech appliances, many developed by American tech giants, offered features like inventory tracking, automated shopping lists, and integration with other smart home devices. Despite its widespread use, the refrigerator faces challenges, especially concerning environmental impact and energy consumption. American innovation continues to address these issues, with ongoing research into more sustainable refrigeration technologies, such as solar-powered and thermoelectric cooling systems.

The invention and evolution of the refrigerator are testaments to American innovation and its global influence. From early ice-making machines to modern smart refrigerators, this journey reflects America's spirit of invention, improvement, and adaptation. The refrigerator has not only reshaped domestic life but has also had a profound impact on food consumption, safety, and culture worldwide, symbolizing the far-reaching effects of a single technological advancement.

WASHING MACHINE

The story of the washing machine begins long before the advent of modern appliances. For centuries, laundry was a laborious task, typically done by hand in streams or washbasins. However, the industrial revolution in America set the stage for a transformation in domestic chores, including laundry. The 18th and 19th centuries saw several American inventors experimenting with devices to ease the burden of washing clothes. In 1782, an American, H. Sidgier, received a patent for a rotating drum washer, which laid the groundwork for future washing machines. This was followed by William Blackstone's invention in 1874, who built a machine as a birthday present for his wife. Blackstone's machine was a hand-operated device that removed clothes' dirt through a board with wooden pegs.

The true revolution in laundry came with the introduction of electricity into American homes. In the early 20th century, Alva J. Fisher, an American inventor, is often credited with inventing the first electrically powered washing machine in 1908, called the "Thor." Manufactured by the Hurley Machine Company of Chicago, the Thor was a drum-type washing machine with a galvanized tub and an electric motor. The washing machine began to significantly impact American domestic life in the 1920s and 1930s. It freed up a considerable amount of time, particularly for women, and became a symbol of modern convenience and efficiency. The washing machine's evolution mirrored the changing role of women in society, as it afforded them more time for work, education, and leisure.

The post-World War II era saw a boom in home appliance ownership in America, including washing machines. This period marked a significant shift, with washing machines becoming a standard appliance in American homes. The American culture of convenience and efficiency began to spread globally, as other countries started to adopt these appliances. Throughout the 20th century, American companies like General Electric, Maytag, and Whirlpool were at the forefront of washing machine innovation. These companies continually improved the design, efficiency, and functionality of washing machines, introducing features like automatic timers, various wash cycles, and advanced spinning technology.

The American innovation in washing machines had a ripple effect across the globe. Countries around the world began to embrace the convenience of washing machines, leading to changes in domestic habits and freeing up time for other activities. The washing machine became an essential tool in the pursuit of modern living standards. As environmental awareness grew towards the late 20th century, American manufacturers began focusing on making washing machines more energy and water-efficient. This led to the development of front-loading machines, which use less water and electricity than top-loading models. The digital age brought further advancements, with American companies integrating smart technology into washing machines. These modern machines offer features like remote control operation, self-diagnosis for repairs, and customization of washing cycles. Despite their benefits, washing machines present challenges, such as water usage, energy consumption, and detergent pollution. American innovation continues to address these issues, with ongoing research into more sustainable and eco-friendly washing technologies.

Looking towards the future, the potential for washing machine technology is vast. Innovations like waterless washing and AI-integrated machines are on the horizon. American companies and startups are exploring ways to revolutionize laundry further, making it more efficient, environmentally friendly, and accessible to all parts of the world. The invention and evolution of the washing machine symbolize American ingenuity and its impact on the world. From alleviating the burden of manual laundry to reshaping domestic life and contributing to environmental sustainability, the washing machine's journey reflects America's spirit of innovation and its far-reaching global influence.

DISPOSABLE DIAPERS

For centuries, diapering was a constant and labor-intensive task. Traditional cloth diapers, while reusable, required frequent washing and were often inefficient at keeping babies dry and comfortable. The journey towards a more practical solution began in the early 20th century, with several attempts to create a disposable diaper. However, it wasn't until the post-World War II era that the disposable diaper as we know it began to take shape.

The real breakthrough in the disposable diaper story came from an American housewife and mother, Marion Donovan. Frustrated with the constant laundry and leaking cloth diapers, Donovan invented the "Boater," a waterproof covering made from a shower curtain, in 1946. This innovation kept cloth diapers from leaking and was a significant step towards more practical diapering solutions.

Marion Donovan didn't stop at the Boater. She continued her work, aiming to create a fully disposable diaper. Her design, which used disposable absorbent material inside a plastic covering, was initially rejected by established manufacturers. However, her persistence paid off when she patented her design in 1951, setting the stage for the disposable diapers we know today. While Donovan laid the groundwork, it was the American company Procter & Gamble (P&G) that brought disposable diapers to the masses. In 1961, P&G introduced Pampers, the first widely successful disposable diaper brand. These early Pampers were a far cry from today's thin and highly absorbent models. They were bulky and used fluff pulp with a rayon top sheet and a plastic backing.

The introduction of disposable diapers had a profound impact on American families. It revolutionized childcare, making it more manageable and freeing up time for parents. For mothers, particularly, disposable diapers offered greater convenience, impacting their ability to participate in the workforce and pursue activities outside of homemaking. The convenience of disposable diapers soon caught on globally. Countries around the world began adopting this American innovation, leading to significant changes in childcare practices. The disposable diaper became a symbol of modern parenting, offering a practical solution to a universal need.

Over the years, disposable diapers have undergone continuous improvement and innovation, much of it driven by American companies. The 1980s and 1990s saw the introduction of superabsorbent polymers, making diapers thinner and more absorbent. Other innovations included resealable tabs, elastic leg bands for better fit, and gender-specific designs. As disposable diapers became ubiquitous, they also raised environmental concerns. The issue of diaper waste – non-biodegradable materials filling landfills – became a growing concern. In response, American companies and startups began exploring more sustainable options, such as biodegradable materials and improved recycling technologies.

Today, the disposable diaper industry is a multi-billion dollar global market. American companies like P&G and Kimberly-Clark (the maker of Huggies) continue to lead, with ongoing innovations focusing on sustainability, skin health, and smart diaper technology – integrating sensors for health monitoring. The future of disposable diapers lies in balancing convenience with environmental responsibility. Ongoing research and innovation in materials science, waste management, and sustainable manufacturing processes are crucial. American ingenuity continues to play a pivotal role in these developments, driving the industry towards more eco-friendly and efficient solutions.

The invention and evolution of disposable diapers are testaments to American innovation and its ability to address everyday challenges effectively. From Marion Donovan's initial invention to the high-tech diapers of today, this journey reflects the American spirit of problem-solving and improvement. The global adoption of disposable diapers illustrates how an American innovation can transform daily life across cultures and continents, highlighting the profound impact of a seemingly simple idea.

VACCINES

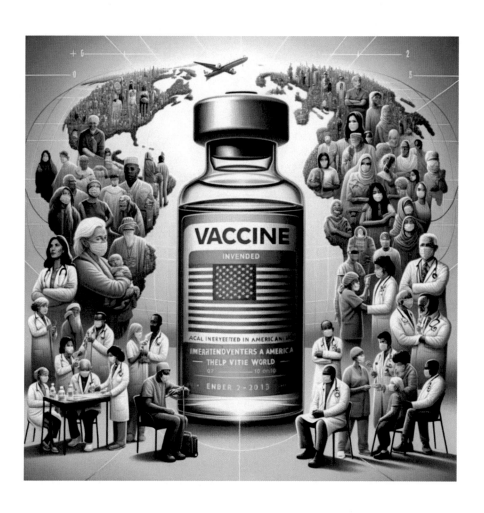

While the concept of vaccination dates back centuries, the modern era of vaccines began with the work of Louis Pasteur in the 19th century. However, it was in the 20th century, particularly in the United States, that vaccine research and development truly accelerated. One of the most dramatic chapters in the history of American vaccines is the development of the polio vaccine. Polio a terrifying disease, causing paralysis and death across the United States and the world. The breakthrough came in the mid-20th century when American medical researcher Jonas Salk developed the first effective polio vaccine.

Salk's work, conducted at the University of Pittsburgh, led to the introduction of the inactivated polio vaccine (IPV) in 1955. This was a monumental moment in medical history. Mass immunization campaigns followed, dramatically reducing polio cases. Salk's colleague, Albert Sabin, furthered this success with the development of the oral polio vaccine (OPV) in the early 1960s, which was easier to administer and became the backbone of global polio eradication efforts.

Smallpox, a disease that had plagued humanity for centuries, was declared eradicated in 1980, a testament to the power of vaccines. While the initial vaccine was not an American invention, the United States played a crucial role in the global eradication campaign. Another milestone in American vaccine history is the development of the measles vaccine. John Enders and colleagues at Boston Children's Hospital developed the first measles vaccine in 1963. Before this vaccine, millions suffered from measles annually. The vaccine has since led to a dramatic decrease in measles cases worldwide. The development of the hepatitis B vaccine was another significant American contribution. In the 1960s, Baruch Blumberg, an American physician, discovered the hepatitis B virus, and in the 1980s, the first hepatitis B vaccines were developed. This vaccine has had a profound impact, particularly in preventing liver diseases and liver cancer globally.

The late 20th and early 21st centuries saw a revolution in vaccine technology with the advent of genetic engineering. This era heralded new types of vaccines, like the recombinant hepatitis B vaccine and the human papillomavirus (HPV) vaccine, which have had significant impacts on public health. In the 21st century, American research and innovation have been crucial in responding to emerging diseases. American laboratories and companies have been at the forefront of developing vaccines for diseases like Ebola and Zika.

The COVID-19 pandemic brought an unprecedented challenge, and American science was once again at the forefront. American companies, in collaboration with international partners, used novel mRNA technology to develop COVID-19 vaccines at a record pace. This achievement not only helped curb the pandemic but also opened new avenues for vaccine research and development. American contributions to vaccine development have had a profound impact on global health. Diseases that were once widespread and deadly are now rare or eradicated. American-led initiatives have also played a significant role in making vaccines accessible in low- and middle-income countries, saving countless lives.

Despite these successes, challenges remain. Vaccine hesitancy, equitable distribution, and the emergence of new pathogens are ongoing issues. The American scientific community continues to work on improving vaccine technologies, including developing universal vaccines for influenza and continuing the quest for an HIV vaccine.

The story of vaccines, heavily influenced by American research and innovation, is one of the greatest success stories in public health. It's a story of overcoming challenges, pushing the boundaries of science, and saving lives. As we look to the future, the legacy of these American innovations continues to inspire and drive progress in protecting global health. This comprehensive story captures the essence of American contributions to vaccine development, highlighting key inventions and their global impact, while also addressing the challenges and the future of vaccinology.

BARCODE

The story of the barcode begins in America in the late 1940s. The idea was first conceived by Bernard Silver and Norman Joseph Woodland, two young graduate students at Drexel Institute of Technology in Philadelphia. Silver overheard a local supermarket manager discussing the challenges of automatically capturing product information at the checkout. Intrigued, Silver and Woodland set out to solve this problem. In 1949, Woodland and Silver developed a system that used light to read product information. Inspired by Morse code, they created a pattern of wide and narrow lines that could be scanned and interpreted as numerical data. This invention, patented in 1952, laid the groundwork for the modern barcode system.

While Woodland and Silver conceptualized the barcode, it wasn't immediately implemented. The technology to read and process the barcode efficiently didn't exist yet. It wasn't until the 1960s and 1970s, with the advancement of laser and computer technology, that their invention could be fully utilized. The pivotal moment in the barcode story came in the early 1970s with the development of the Universal Product Code (UPC). George Laurer, an engineer at IBM, played a crucial role in developing the UPC system, which standardized the barcode format and made it universally readable. In 1974, the first product with a UPC barcode, a pack of Wrigley's gum, was scanned at a supermarket in Troy, Ohio. This event marked the beginning of a new era in retail and supply chain management.

The adoption of the barcode system revolutionized the retail industry in America and then globally. Barcodes increased the efficiency of checkout processes, improved inventory management, and reduced human errors. This technology enabled retailers to keep better track of their products and understand consumer purchasing patterns, leading to more efficient and customer-focused practices.

The success of the barcode in the United States led to its rapid adoption worldwide. The International Article Numbering Association (now GS1), founded in 1977, played a key role in standardizing barcodes internationally, ensuring that products could be tracked and scanned anywhere in the world. The barcode's impact extended far beyond retail. It transformed supply chain management across various industries. By enabling precise tracking of goods, barcodes optimized inventory levels, reduced costs, and improved delivery times. This efficiency was critical in the era of globalization, as products increasingly crossed international borders.

The evolution of barcode technology continued with the development of two-dimensional (2D) barcodes, like QR codes, which can store more information than traditional barcodes. This advancement expanded the use of barcode technology into new areas, including marketing, where QR codes link directly to digital content on smartphones. In the United States, barcode technology has been instrumental in healthcare, improving patient safety and medication management. Barcodes on medication packages and hospital wristbands ensure that patients receive the correct medication and dosage. The story of the barcode is a testament to the culture of innovation in America. It reflects a society that encourages creative problem-solving and supports technological advancements. The journey from Woodland and Silver's initial idea to the global implementation of the barcode system underscores the American spirit of entrepreneurship and collaboration between different sectors, including academia, industry, and government. More recently, barcode technology has been used in environmental efforts, such as tracking the sustainability of products and monitoring the supply chains of environmentally sensitive goods, such as seafood and timber.

Looking to the future, the potential of barcode technology is vast. Developments in data storage, scanning technologies, and integration with internet-of-things devices could lead to even more innovative uses of barcodes. American companies and research institutions continue to be at the forefront of exploring these possibilities. The invention and development of the barcode is a story that encapsulates the essence of American innovation. A simple yet groundbreaking idea transformed the way the world conducts business, manages products, and processes information. As we look towards the future, the legacy of the barcode, rooted in American ingenuity, continues to influence industries worldwide, demonstrating the far-reaching impact of a visionary concept.

BAND-AID

The American story of the Band-Aid begins in the 1920s in New Jersey, where a young cotton buyer named Earle Dickson lived with his wife, Josephine. Josephine was prone to kitchen accidents, often cutting and burning herself while cooking. The couple's plight was common in households everywhere: minor injuries were a nuisance to treat effectively with the available first-aid supplies, typically gauze and adhesive tape, which were cumbersome and not always easy to apply one-handed.

Dickson, an employee at Johnson & Johnson, a company already well-established in the field of surgical dressings, took it upon himself to find a solution for his wife's frequent injuries. His innovation was elegantly simple: he placed small pieces of sterile gauze at intervals along a strip of adhesive tape, then covered the setup with crinoline to keep it sterile. This design allowed his wife to dress her wounds quickly and efficiently, without assistance.

Recognizing the potential of his invention, Dickson presented it to his employer. Johnson & Johnson saw the promise in Dickson's creation and began producing Band-Aids in 1920. Initially, they were handmade and not an immediate success, as they were relatively unknown and a bit ahead of their time. The Band-Aid's popularity soared with a crucial shift in strategy. In 1924, Johnson & Johnson introduced machine-made Band-Aids and began sterilizing them, addressing concerns about infection and making the product more appealing to the general public. But it was the company's marketing strategy that truly turned the Band-Aid into a household name.

Johnson & Johnson cleverly distributed free Band-Aids to Boy Scout troops across America, embedding the product into

American culture. They also targeted marketing efforts at children, recognizing that parents were more likely to buy Band-Aids for their children's frequent minor injuries. World War II was a turning point for the Band-Aid. Johnson & Johnson supplied vast quantities of Band-Aids to the American military, which helped introduce the product to a global audience. Soldiers who used Band-Aids on the battlefield brought their appreciation for the product back home, spreading its popularity both in the U.S. and abroad.

Over the years, the Band-Aid evolved. The introduction of vinyl and other plastic materials in the 1950s allowed for waterproof and more durable varieties. The product line expanded to include a range of sizes, colors, and even antibiotic-impregnated dressings. These innovations kept the Band-Aid relevant and essential in changing times. The Band-Aid's impact goes beyond its practical use. It has become a cultural icon, symbolizing care, protection, and the simple yet effective solutions that are often the hallmark of American innovation. The phrase "Band-Aid solution" entered the lexicon, denoting a temporary or quick-fix solution to a problem. The Band-Aid's story is a testament to American ingenuity and its ability to address everyday problems in a simple and effective way. From Earle Dickson's initial idea to its global proliferation, the Band-Aid has become an indispensable part of medical care and a symbol of quick, reliable healing. Its journey from a New Jersey home to households worldwide highlights how a small innovation can have a vast and enduring impact.

SAFETY RAZOR

For centuries, shaving was a challenging and often dangerous task. The straight razor, the primary tool for shaving, required skill and care to use safely. It was not only easy to cut oneself, but the straight razor also needed regular sharpening and maintenance. This cumbersome process called for an easier, safer way to shave, a need that sparked a significant innovation in the early 20th century.

The pivotal figure in the story of the safety razor is King Camp Gillette, an American businessman and inventor. In the late 1800s, Gillette conceived the idea of a safety razor with disposable blades. His vision was revolutionary: a razor that was safe, convenient, and easy to use, with a disposable blade that didn't require sharpening. Gillette's idea was initially met with skepticism. Many believed it was impossible to manufacture a disposable blade cheaply enough to be practical. However, Gillette persisted, and in 1901, with the help of William Nickerson, an MIT-trained engineer, he founded the American Safety Razor Company (later renamed Gillette Safety Razor Company).

In 1903, Gillette introduced the first safety razor to the market. It featured a double-edged blade that could be used on both sides, housed in a head that provided a protective guard between the blade and the skin. This design significantly reduced the risk of cuts, revolutionizing the shaving experience. The early years were challenging, with slow sales. However, Gillette's big break came with the outbreak of World War I. The U.S. government ordered millions of Gillette razors and blades to provide to American troops. This massive order not only boosted Gillette's business but also introduced the safety razor to countless American soldiers.

The distribution of Gillette safety razors to the military had far-reaching consequences. Soldiers who used these razors during the war brought their preference for them back home, dramatically boosting demand in the post-war years. This exposure effectively popularized the safety razor, making it a staple in households across America and eventually around the world. Over the years, the safety razor underwent numerous innovations. The 1930s saw the introduction of the twist-to-open design, which made changing blades easier. In the 1970s, the disposable razor was introduced, offering even greater convenience by eliminating the need to change blades. These innovations, many of which were driven by American companies like Gillette, continued to shape the shaving industry.

The popularity of the safety razor soon spread beyond the United States. Companies began exporting these razors worldwide, making them a common feature in bathrooms across the globe. The safety razor became synonymous with modern grooming, appreciated for its ease of use, safety, and affordability. The safety razor's impact extended beyond its practical use. It symbolized a shift towards convenience and consumerism in American culture, a trend that was rapidly spreading worldwide. The safety razor became a symbol of the modern man, sleek and efficient, paralleling the fast-paced progress of the 20th century.

In recent years, there has been a growing awareness of the environmental impact of disposable razors. This consciousness has led to a resurgence in the popularity of traditional safety razors, which are seen as more sustainable and eco-friendly. Modern iterations combine the classic design with contemporary materials and features, reflecting a blend of tradition and innovation. The story of the safety razor is a testament to American ingenuity and its capacity to transform everyday life. King C. Gillette's vision and perseverance not only changed the way men shave but also sparked an industry that has had a lasting global impact. From its humble beginnings in the United States to its worldwide presence, the safety razor remains an enduring symbol of practical innovation and convenience.

HEARING AID

The invention of the hearing aid is a remarkable story of American innovation, a journey of discovery and refinement that has profoundly impacted lives globally. The history of hearing aids dates back centuries, with various rudimentary devices like ear trumpets used to amplify sound. However, these early solutions were bulky, ineffective, and often socially stigmatizing. The real transformation in hearing assistance began with the advent of electricity and the telephone in the 19th century, laying the groundwork for the modern hearing aid.

The invention of the telephone by Alexander Graham Bell, a Scots-American inventor, was a pivotal moment in hearing aid history. Bell's mother and wife were both hearing impaired, which drove his interest in acoustic science. The telephone's technology – converting sound into electrical signals and then back into sound – was directly applicable to developing an effective hearing aid. The first practical electrical hearing aids, developed in the early 20th century, used carbon microphones, similar to early telephones. These carbon hearing aids amplified sounds but were bulky and had limited frequency range. Despite these limitations, they represented a significant step forward in hearing technology.

The next significant advancement came in the 1920s with the introduction of the vacuum tube hearing aid. This technology, developed by American companies, including Western Electric, significantly improved sound amplification and quality. Vacuum tube hearing aids were smaller, more effective, and marked the beginning of hearing aids as we know them today. The mid-20th century saw a push towards miniaturization and portability in hearing aids, driven by advancements in electronics. The invention of the transistor in 1947, a critical development by Bell Laboratories in the United

States, revolutionized many electronic devices, including hearing aids. Transistor-based hearing aids were smaller, more energy-efficient, and provided better sound quality than their predecessors.

In the 1960s, American innovation led to the development of behind-the-ear (BTE) hearing aids. These devices housed all their electronic components in a small case worn behind the ear, connected to an ear mold that fit inside the ear. BTE hearing aids were more discreet and comfortable, offering greater acceptance and a better quality of life for users. The digital revolution in the late 20th century marked a new era in hearing aid technology. Digital hearing aids, developed through groundbreaking work in American technology firms and universities, offered unprecedented sound quality, customization, and features. They could be programmed to match individual hearing loss patterns and reduce background noise, significantly improving user experience. The 21st century saw further innovations with the integration of wireless technology. Modern hearing aids, many developed by American companies, can now connect wirelessly to smartphones, TVs, and other devices, enhancing communication capabilities for users. This connectivity represents a significant shift in how hearing aids are perceived and used, integrating them seamlessly into the digital world.

The American-led innovations in hearing aid technology have had a global impact. Millions of people around the world benefit from improved hearing, thanks to devices that are more effective, comfortable, and accessible than ever before. These advancements have not only enhanced the ability to hear but also reduced the social stigma associated with hearing loss. Despite these advancements, challenges remain in making hearing aids more accessible and affordable globally.

Future directions in hearing aid development, driven by ongoing American innovation, include further miniaturization, more sophisticated sound processing, and integrating health monitoring features.

The invention and evolution of the hearing aid is a testament to American innovation's power to change lives. From the early electrical devices to today's digital, wireless hearing aids, this journey reflects a continuous pursuit of improving the human experience. As a tool that has opened up the world of sound to millions, the hearing aid stands as a symbol of how American ingenuity and perseverance can create solutions with a lasting and profound global impact.

DUCT TAPE

The invention of duct tape is a quintessential example of American ingenuity, a story of innovation born out of necessity that has since become a staple in toolboxes around the world.. The story of duct tape begins in the early 1940s during World War II. An American factory worker and mother of two Navy sailors, Vesta Stoudt, first conceptualized the idea. Stoudt worked in a munitions plant where she noticed a critical flaw in the packaging of ammunition boxes. These boxes were sealed with paper tape and wax, which was not water-resistant and difficult to open quickly in the heat of battle. Concerned for the safety of soldiers, including her sons, Stoudt envisioned a strong, waterproof, and easy-to-use adhesive solution.

Stoudt sent a letter detailing her idea to President Franklin D. Roosevelt, who recognized the potential of her suggestion and passed it on to the War Production Board. The task of developing this new tape was given to Johnson & Johnson, a company already renowned for its adhesive bandages. Led by their Revolite division, which specialized in making treated fabrics, Johnson & Johnson developed a new adhesive tape made from a durable duck cloth fabric. This innovative tape was not only strong and waterproof but also could be torn by hand and used to securely seal ammunition boxes while being easy to open when needed. This new tape, initially referred to as "duck tape" due to its duck cloth backing, was an instant success in the military. Soldiers soon discovered its versatility, using it for everything from repairing equipment to waterproofing boots. It was silver-gray in color and incredibly adhesive, leading to its nickname "duck tape" among the soldiers.

After World War II, duct tape transitioned from military use to civilian life. In the post-war housing boom, America saw the widespread adoption of central heating and air conditioning systems. Duct tape, already known for its strength and versatility, became the go-to tool for connecting heating and air conditioning ducts. This new application led to the evolution of its name from "duck tape" to "duct tape." American companies began manufacturing duct tape for commercial use. Brands like 3M and Duck Brand started producing their own versions of duct tape, improving its adhesive properties and expanding its use in various industries. The tape was now available in different colors and sizes, catering to a wide range of needs.

Duct tape became a symbol of the American "can-do" attitude. Its ability to fix a multitude of problems quickly and effectively made it a favorite among DIY enthusiasts, handymen, and even astronauts. NASA famously used duct tape during the Apollo 13 mission to repair equipment, showcasing its reliability even in the most challenging situations. The popularity of duct tape spread globally. Its ease of use and effectiveness made it popular in various settings, from households to industrial environments. Duct tape was used for quick repairs, temporary fixes, and even in artistic and creative projects. Over the years, duct tape has seen numerous innovations. Manufacturers have developed varieties that offer greater durability, different colors and patterns, and tapes designed for specific applications, such as electrical work or outdoor use. The story of duct tape is a testament to the power of practical, everyday innovation. From its humble beginnings in a World War II factory to its status as a ubiquitous tool in homes and industries worldwide, duct tape embodies the spirit of American ingenuity. Its journey from a simple idea to a global phenomenon highlights how an invention born out of necessity can become an indispensable part of daily life across cultures and continents.

AIR CONDITIONER

The air conditioner was invented in the United States. The inventor of the modern air conditioner is Willis Haviland Carrier. He designed the first modern air conditioning system in 1902. The late 19th and early 20th centuries in America were a period of rapid industrialization. Factories and print shops were burgeoning, but they faced a significant challenge: the sweltering heat of the American summers. This heat was more than just a discomfort; it affected productivity, spoiled goods, and made working conditions intolerable.

Willis Haviland Carrier, born in 1876 in Angola, New York, was a gifted engineer who embarked on a mission that would eventually revolutionize the way we live. In 1902, working for the Buffalo Forge Company, Carrier was tasked with solving a humidity problem that was causing magazine pages to wrinkle at the Sackett-Wilhelms Lithographing and Publishing Company in Brooklyn. Carrier's solution was not just to cool the air but to control its humidity. He invented a system that passed air over coils cooled with water, both lowering the air's temperature and removing moisture to control humidity. This invention, often referred to as the "Apparatus for Treating Air" (patented in 1906), was the world's first modern air conditioning system.

Carrier's invention initially found its application in industrial settings, particularly in industries where humidity control was crucial, like textile mills and pharmaceutical plants. However, it wasn't long before the potential of air conditioning for improving general comfort was realized. In the 1920s, the technology began to spread to public spaces. The Rivoli Theater in Times Square, New York, was one of the first public places to install air conditioning in 1925, leading to the summer blockbuster phenomenon in the film industry, as people flocked to air-conditioned theaters during hot summer months. The real boom in air conditioning came post-World War II. With the American economy booming and technological advancements made during the war, air conditioning began to be seen not as a luxury but as a necessity. The development of compact, efficient, and affordable window air conditioners meant that, for the first time, it was feasible to install air conditioning in homes.

This change had a profound impact on American society and architecture. It allowed for the population boom in the Sun Belt states like Florida, Texas, and Arizona. Modern skyscrapers, with their sealed windows and centralized air conditioning systems, became possible, reshaping the skylines of cities around the world. The impact of air conditioning was not confined to the United States. Post-World War II, as American culture and technology spread globally, so did air conditioning. Countries in the Middle East, Asia, and other parts of the world with hot climates began to adopt air conditioning, changing living and working patterns.

The widespread adoption of air conditioning brought with it environmental concerns. The use of chlorofluorocarbons (CFCs) in air conditioning contributed to the depletion of the ozone layer, leading to international treaties like the Montreal Protocol, which phased out their use. This challenge spurred further innovation. New, more environmentally friendly refrigerants were developed, and advances in technology made air conditioners more energy-efficient and less harmful to the environment.

Today, air conditioning technology continues to evolve. Smart thermostats, AI-driven efficiency optimization, and sustainable practices are at the forefront of current developments in the industry. The focus has shifted to not just providing comfort but doing so in an environmentally responsible and sustainable way. The story of air conditioning is a quintessential American tale of innovation addressing a practical problem, with far-reaching consequences that changed the world. It reshaped industries, influenced global migration and urban development, and significantly improved the quality of life. As we look to the future, the legacy of Willis Haviland Carrier and his invention continues to evolve, reminding us of the ongoing journey of human ingenuity and its global impact. This narrative, weaving together historical developments, technological advancements, and societal changes, provides a comprehensive view of the invention of air conditioning and its global impact, emphasizing the American roots and innovative spirit that drove this transformative invention.

ROLLER COASTER

The roots of the roller coaster trace back to the ice slides popular in Russia in the 16th and 17th centuries. These large, sloping structures, built of ice and snow, provided a form of winter amusement for the Russian aristocracy. The concept evolved in France in the early 1800s, where the French added wheels to carts and tracks to create the "Les Montagnes Russes" or Russian Mountains.

The real transformation that led to the modern roller coaster, however, began in America. LaMarcus Adna Thompson, often hailed as the "Father of the Gravity Ride," played a pivotal role. In 1884, Thompson patented his design of the Gravity Switchback Railway, which he debuted at Coney Island, New York. This ride, while modest by today's standards, was a sensation of its time. It featured a series of undulating tracks on which cars glided along powered by gravity.

Thompson's invention sparked a roller coaster craze in America. It coincided with the country's burgeoning fascination with technology and innovation in the late 19th and early 20th centuries. The roller coaster became a symbol of America's industrial prowess and inventiveness. Following Thompson's success, entrepreneurs and engineers across the country began creating more elaborate and thrilling designs. These early roller coasters were primarily wooden structures, renowned for their rickety and exhilarating rides. The 1920s saw the golden age of roller coasters in America. Legendary designers like John A. Miller and Harry Traver emerged, pushing the boundaries of roller coaster engineering. Miller, in particular, was a prolific designer who introduced many features that became staples in roller coaster design, including the underfriction, or "upstop," wheel which allowed for more ambitious and safer designs.

The Great Depression and World War II brought a temporary decline in the roller coaster industry. Economic hardships and material rationing led to fewer new roller coasters being built, and many existing ones fell into disrepair. Post-World War II America experienced a revival in the roller coaster industry. This era saw the transition from traditional wooden roller coasters to steel roller coasters. The introduction of steel allowed for even more innovative designs, including loops, corkscrews, and higher, faster tracks. The shift to steel roller coasters marked a new era of innovation. In 1959, Disneyland introduced the Matterhorn Bobsleds, the first tubular steel roller coaster, setting off a wave of steel coaster constructions. The flexibility of steel allowed for more intricate designs and smoother rides, revolutionizing the roller coaster experience. One of the most significant advancements in roller coaster design was the introduction of the vertical loop. The first modern vertical looping coaster, Corkscrew, was unveiled at Knott's Berry Farm in California in 1975. This ushered in an age of roller coasters with multiple inversions and complex layouts.

Today's roller coasters are marvels of engineering and technology. Computer-aided design (CAD) software, precise engineering, and advanced safety technologies have led to the creation of some of the tallest, fastest, and most thrilling rides in the world. The American innovation in roller coaster design has had a global impact. Countries around the world have embraced the roller coaster, building their own colossal rides and theme parks. The roller coaster has become a universal symbol of fun and excitement, transcending cultural and language barriers.

The roller coaster's story is a testament to American creativity and its capacity to bring joy and excitement to people worldwide. From Thompson's Gravity Switchback Railway to the steel giants of today, the roller coaster continues to evolve, each new design pushing the limits of imagination and engineering. As a symbol of fun and innovation, the roller coaster stands as a proud representation of American ingenuity and its enduring global appeal.

GPS

Navigating the World: The American Invention of GPS

Imagine a world without the convenience of GPS navigation—a world where getting lost was a common occurrence and finding your way in unfamiliar places was a challenge. This chapter explores the remarkable story behind the creation of GPS (Global Positioning System), its American origins, and its profound impact on navigation worldwide.

Before GPS, navigating across land, sea, and air relied on a combination of maps, compasses, and celestial observations. While these methods were sufficient for centuries, they had limitations, especially when it came to pinpoint accuracy and real-time tracking. The journey toward GPS began in the early 1960s with the U.S. Department of Defense. They envisioned a revolutionary navigation system that could provide precise positioning information to military forces and enable accurate missile guidance. Thus, the concept of GPS was born.

The development of GPS involved a team of American engineers and scientists who tackled the complex challenges of sending and receiving signals from satellites in orbit. Their groundbreaking work laid the foundation for a navigation system that would soon change the world. In 1978, the U.S. launched its first navigation satellite, Navstar 1, marking the beginning of the GPS constellation. Over the years, more satellites were launched, forming a network that covered the entire globe.

Initially developed for military purposes, GPS technology quickly demonstrated its potential for civilian applications. In 1983, after the Korean Air Flight 007 incident, President Ronald Reagan announced that GPS would be made available for civilian use, heralding a new era of navigation. GPS relies on a constellation of satellites orbiting Earth, each emitting precise timing signals. By triangulating signals from multiple satellites, GPS receivers can calculate their exact position, elevation, and velocity. This remarkable technology brought unprecedented accuracy to navigation. The introduction of GPS revolutionized navigation in various sectors. It became an invaluable tool for aviation, maritime, transportation, agriculture, surveying, and emergency response. GPS made it possible to navigate with pinpoint accuracy, even in the most challenging environments.

GPS is a quintessential American innovation that has had a profound impact on navigation worldwide. Its origins in the U.S. represent the country's commitment to pushing the boundaries of technology and improving the lives of people around the globe. Imagine a world without GPS. Maritime navigation would rely more on sextants and charts, aviation would face greater challenges in adverse weather conditions, and land-based travel would be less efficient. The absence of GPS would affect everything from logistics and agriculture to search and rescue missions.

GPS has transformed agriculture, allowing farmers to apply fertilizers, pesticides, and irrigation with precision. This technology not only increases crop yields but also reduces environmental impact by minimizing overuse of resources. GPS has become a lifeline in emergency response situations. Search and rescue teams, paramedics, and disaster relief organizations rely on GPS to locate and assist those in need quickly and efficiently. GPS has seamlessly integrated into our daily lives. It guides us to our destinations, helps us find nearby businesses, and provides real-time traffic information. From smartphones to in-car navigation systems, GPS has become an essential companion. As technology continues to advance, GPS is evolving as well. The system is being upgraded to improve accuracy, reliability, and availability. Augmented reality and autonomous vehicles are among the exciting fields that rely heavily on GPS technology.

The roots of GPS can be traced back to the early days of space exploration and the Cold War. In the 1950s, the United States and the Soviet Union were engaged in a race to conquer space. This era of competition and technological advancement led to the launch of the first artificial satellites, Sputnik by the Soviets in 1957, and Explorer 1 by the United States in 1958. These events marked the beginning of the space age and set the stage for the development of GPS. The U.S. Department of Defense recognized the need for a reliable and accurate navigation system for its military operations. Traditional methods of navigation, such as celestial observations and dead reckoning, were often imprecise and inadequate for modern warfare. The military needed a system that could provide real-time positioning information, especially for missile guidance and targeting.

In the early 1960s, the concept of a global navigation system using satellites was proposed. The U.S. Navy played a significant role in advancing this idea. The Navy's Navigation Satellite System (NSS) project aimed to develop a satellite-based navigation system that could provide accurate positioning data to ships and submarines. This project laid the groundwork for the eventual development of GPS. However, the true breakthrough came with the launch of the first GPS satellite, Navstar 1, on February 22, 1978. This marked the beginning of the GPS constellation, which would eventually consist of a network of 24 satellites. These satellites were strategically positioned in orbit around the Earth, forming a reliable and precise global positioning system. The development of GPS involved a team of dedicated American engineers and scientists who faced numerous technical challenges. One of the primary challenges was the synchronization of signals from multiple satellites. To determine a receiver's position accurately, it needed to receive signals from at least four satellites simultaneously. Achieving this synchronization was no small feat, but it was essential for the system's success.

Another critical aspect was the development of highly accurate atomic clocks on board the satellites. Precise timing was crucial for calculating the distance between the satellites and the receiver on the ground. Any deviation in timing could result in significant errors in the calculated position. Overcoming these challenges required innovation and collaboration. The U.S. government, in partnership with private industry, invested heavily in research and development to make GPS a reality. The development of the system's control segment, ground stations, and user equipment was equally important in ensuring the success of GPS.

The invention of GPS, with its American origins, symbolizes a commitment to innovation and progress. It is a testament to the American spirit of exploration and a reminder that even the most complex challenges can be overcome with determination and ingenuity. The story of GPS is one of American innovation that has shaped the way the world navigates. From its military origins to its civilian applications, GPS has become an indispensable tool that has improved the efficiency, safety, and convenience of navigation across the globe. As we reflect on the impact of GPS, we are reminded that even in our increasingly interconnected world, American inventions like GPS continue to play a pivotal role in shaping the future and empowering individuals and societies worldwide.

ROBOTS

The concept of robots has been around for centuries, found in myths and folklore. However, the journey towards actual robotic development began in the early 20th century. While robots are not solely an American invention, as early developments occurred worldwide, American innovation has significantly shaped the field.

The term "robot" was first introduced in a 1920 Czech play, but it was in the United States where much of the early development in robotics took place. The 1950s and 1960s marked the beginning of modern robotics. American companies and researchers started experimenting with machines that could perform tasks automatically. One of the first robots, Unimate, was developed in the U.S. by George Devol and Joseph Engelberger and installed in a General Motors factory in 1961. It was a programmable arm that could perform repetitive tasks - a breakthrough in industrial automation.

Initially, robots were mainly used in car manufacturing to handle heavy lifting and dangerous tasks. However, as technology advanced, so did the capabilities of robots. American researchers and companies began developing robots for various uses, from medical surgeries to space exploration. In the 1980s and 1990s, robots began to enter the public consciousness more significantly. American movies and television shows often featured robots, sparking public interest and imagination about their possibilities. The late 20th and early 21st centuries saw a technology boom, and robotics was at the forefront of this revolution. American universities and tech companies made significant strides in advancing robotic technology. Robots became smarter, smaller, and more affordable, leading to a wider range of applications.

A key development in the field of robotics has been the advancement in artificial intelligence (AI) and machine learning, much of it driven by American innovation. These technologies allow robots to learn from experiences and adapt to new tasks, making them more versatile and efficient. This advancement has led to robots that can perform complex tasks, from driving cars to providing customer service.

Today, robots are part of everyday life in many parts of the world. American companies like iRobot have popularized robotic vacuum cleaners in households. Robots are also used in hospitals for surgeries and in pharmacies to dispense medication, improving precision and efficiency. Another area where American technology has made a significant impact is in the development of drones. Initially used for military purposes, drones have found their way into commercial and recreational use. They're used for everything from film making to agriculture, showcasing the versatility of robotic technology. In the U.S., robotics has also become a vital part of education and research. American educational institutions are at the forefront of robotics research, pushing the boundaries of what robots can do. They are also incorporating robotics into their curriculum, preparing students for a future where robotics will be increasingly important.

The influence of American robotics can be felt worldwide. American-made robots and robotic technology are used globally in various industries, impacting economies and labor markets. The U.S. has also been a leader in setting standards for robotic technology and its ethical use, influencing global policies and practices. Looking to the future, the possibilities of robotics are boundless. Robots could play a key role in addressing global challenges, from environmental conservation to healthcare. The ongoing research in American institutions and companies is pushing towards more sophisticated robots that could potentially revolutionize how we live and work.

However, with these advancements come challenges and ethical considerations. Issues like job displacement, privacy, and safety are part of the ongoing conversation about the future of robotics. American policymakers and technologists are actively engaged in addressing these concerns.

The story of robotics is an ongoing tale of American innovation, with a significant global impact. From industrial automation to personal assistants, robots have changed how we work and live. As American technology continues to lead in the field of robotics, the future holds exciting possibilities, promising to bring even more revolutionary changes to our world.

SOLAR PANELS

The story of solar panels begins in the 19th century. While the concept of harnessing solar energy dates back to ancient times, it was not until 1839 that the French physicist Edmond Becquerel discovered the photovoltaic effect, which forms the basis of solar panels. However, it was American inventors and scientists who significantly advanced this technology.

The modern solar panel as we know it today was born in 1954 at Bell Labs in the United States. Researchers Daryl Chapin, Calvin Fuller, and Gerald Pearson developed the first practical silicon solar cell, which was capable of converting enough sunlight into energy to run everyday electrical equipment. This groundbreaking invention marked the first step in making solar energy a viable alternative to fossil fuels. Initially, solar panels were prohibitively expensive and primarily used in space applications, powering satellites in Earth's orbit. The American space program played a crucial role in advancing solar technology, as NASA used solar energy to power space crafts and satellites, proving the reliability and durability of this technology.

Throughout the 20th century, American researchers and companies continued to refine solar technology, making it more efficient and affordable. Key advancements included the development of polycrystalline silicon cells in the 1980s, which offered a cheaper alternative to the more expensive monocrystalline cells. In the 1990s and 2000s, American innovation in solar technology gathered pace. The U.S. Department of Energy and various private companies invested heavily in research, leading to more efficient photovoltaic cells and the development of thin-film solar cells, which opened up new possibilities for integrating solar panels into a variety of materials and surfaces.

Today, solar energy is one of the fastest-growing sources of renewable energy in the world, and the United States has been a leader in this growth. American companies like First Solar and SunPower are at the forefront of solar panel technology, continually pushing the boundaries of efficiency and affordability. Solar panels are now a common sight in many American homes, businesses, and even in large-scale solar farms that generate megawatts of electricity. The U.S. government's support through tax incentives and subsidies has also played a crucial role in the adoption of solar energy.

The influence of American solar technology extends far beyond its borders. Countries around the world are adopting solar energy, using technologies developed in the United States. American companies are also leading international projects, bringing solar power to developing countries where it can have a profound impact on energy access and sustainability. The shift towards solar energy is a critical component in addressing global environmental challenges. Solar panels produce clean, renewable energy, reducing reliance on fossil fuels and cutting greenhouse gas emissions. The American leadership in solar technology is thus not just an economic or technological success story, but also an essential part of the global response to climate change.

Looking to the future, the possibilities for solar energy are boundless. American researchers are at the forefront of developing new solar technologies, such as solar skins that can seamlessly integrate into building materials, and floating solar farms that can be deployed on reservoirs and lakes, conserving land and water simultaneously. There is also ongoing research into improving energy storage technologies, such as batteries, which is crucial for making solar energy reliable around the clock. The United States is leading the way in developing these next-generation storage solutions.

Despite its tremendous potential, solar energy still faces challenges. These include improving the efficiency of solar panels, reducing the costs of installation, and addressing environmental concerns related to solar panel manufacturing and recycling. The story of solar panels is a testament to American innovation and its global impact. It's a story that reflects not just a technological revolution but a shift towards a more sustainable and environmentally conscious future. As American technology continues to drive advancements in solar energy, the potential for a positive impact on the world and on future generations is immense.

HOLLYWOOD

In the world of entertainment, Hollywood stands as a towering symbol of glamour, storytelling, and cultural influence. Nestled in the hills of Los Angeles, California, Hollywood has not just been a place but a phenomenon that has shaped the global entertainment landscape. From the silent era to the digital age, its history is a tapestry of ambition, creativity, and the American dream.

Hollywood's story is intrinsically linked to America's. It started as a small community and rapidly became the home of the burgeoning film industry. The movies produced here didn't just entertain; they reflected and propagated American values, lifestyles, and ideals. The influence of Hollywood films extended beyond American borders, captivating audiences worldwide and making an indelible mark on global culture.

The Golden Age of Hollywood

The Golden Age of Hollywood, spanning from the late 1920s to the early 1960s, was a period of unparalleled success and influence. This era witnessed the emergence of major studios like MGM, Warner Bros., and Paramount, which became synonymous with cinematic excellence. Stars like Clark Gable, Marilyn Monroe, and Humphrey Bogart became household names, embodying the allure and glamour of American life. During this time, Hollywood movies were more than just entertainment; they were a window into American culture. Iconic films like "Gone with the Wind" and "Casablanca" not only showcased cinematic brilliance but also mirrored American values of heroism, freedom, and the pursuit of happiness. The narratives often revolved around quintessentially American experiences, portraying the country as a land of opportunity and dreams.

The global impact of these films was profound. They

introduced the world to American fashion trends, social norms, and language idioms. Hollywood's portrayal of the American lifestyle captivated international audiences, shaping their perceptions of the United States. This cultural export extended America's influence far beyond its borders, establishing Hollywood as a powerhouse in global entertainment. Furthermore, Hollywood's Golden Age set the foundation for America's soft power. The films served as cultural ambassadors, disseminating American ideologies and values worldwide. They played a crucial role in positioning the United States not just as a political and economic giant, but also as a cultural trendsetter.

Hollywood's Evolution and Global Influence

As the 20th century progressed, Hollywood continuously evolved, adapting to changing tastes, technologies, and societal norms. The advent of color film, the transition to sound, and the rise of special effects were just a few milestones that kept Hollywood at the forefront of global cinema. This period also saw the emergence of new genres, from captivating musicals to thought-provoking dramas, reflecting the diversity and complexity of American society.

Hollywood's influence during this time was not just limited to the realm of entertainment. It shaped global fashion, as audiences emulated the styles of their favorite stars. Phrases and dialogues from popular films became part of everyday language in various countries, indicating the pervasive impact of Hollywood's storytelling. Moreover, Hollywood began to incorporate diverse cultural elements into its films. This not only enriched the storytelling but also resonated with international audiences, fostering a sense of global connection. Movies like "The Sound of Music" and "West Side Story" transcended cultural barriers, showcasing universal themes of love, conflict, and aspiration.

However, Hollywood's portrayal of American life also led to a certain homogenization of global culture, where American ideals and lifestyles were often idealized. This phenomenon raised questions about cultural dominance and the representation of non-American stories in mainstream cinema. Despite these critiques, Hollywood's role in shaping global popular culture remained undeniable. Its ability to captivate audiences across different continents cemented its status as a cultural powerhouse, influencing everything from fashion and music to societal values and norms.

The Power of Storytelling

Hollywood's greatest influence lies in its storytelling. The narratives spun by Hollywood have not only entertained but also shaped perceptions and dreams. These stories, while diverse in genre and style, often carry a distinctly American flavor – a blend of optimism, individualism, and a sense of destiny.

Films like "The Godfather" and "Forrest Gump" did more than just captivate audiences; they offered insights into American life, values, and history. These stories, though fictional, painted vivid pictures of the American experience, from the struggles for success to the complexities of the American Dream. They spoke of resilience in the face of adversity, a theme that resonates deeply with the American spirit.

Moreover, Hollywood's influence extended to its portrayal of heroes and villains, shaping global perceptions of right and wrong, justice, and morality. Superhero films, a genre that has seen immense popularity in recent years, epitomize this. Characters like Superman and Captain America are not just superheroes; they are embodiments of American ideals and principles.

The impact of these narratives is profound and far-reaching. They have the power to inspire, to change attitudes, and even to influence societal norms and politics. Hollywood stories have often served as catalysts for discussion on critical issues like race, gender, and equality, reflecting and sometimes leading societal change.

However, this power also comes with responsibility. Hollywood has been criticized for its lack of diversity and representation, and there is a growing call for stories that more accurately reflect the world's diversity. This is not just a matter of fairness or inclusion; it's about enriching the tapestry of Hollywood storytelling with a multitude of voices and perspectives.

The Golden Stage: Competition and Collaboration

As Hollywood entered the 21st century, the landscape of global cinema began to shift. The rise of film industries in countries like India, China, and South Korea marked the beginning of a more diverse and competitive global market. These industries, each with their unique storytelling styles and cultural nuances, started to challenge Hollywood's dominance and offered audiences around the world alternative perspectives and narratives.

This emergence of global cinema has been a testament to the universal power of storytelling. Films like South Korea's "Parasite" and India's "Baahubali" series have not only been commercial successes but also critically acclaimed, winning hearts and awards worldwide. These movies proved that a film doesn't have to be from Hollywood to captivate a global audience. They offered fresh narratives, rich in their cultural heritage, yet universal in their appeal.

The success of these non-American films raises an important question: Is it possible for a movie not made in America to be the best and most-watched in the world? The answer is increasingly leaning towards yes. The global film landscape is becoming more inclusive, with international films gaining recognition and popularity. The ingredients for a globally successful film – a compelling story, emotional resonance, and universal themes – are not exclusive to Hollywood.

The potential for collaboration is immense. Hollywood has started to embrace this diversity, often collaborating with international studios and artists. This not only enhances the quality of the films but also helps in bridging cultural gaps, creating a more interconnected and empathetic global community.

Future of Storytelling

Hollywood's journey through the annals of entertainment history is a testament to the power of cinema in shaping and reflecting culture. From the silent films of the early 20th century to the digital blockbusters of today, Hollywood has not only entertained the world but also influenced it in myriad ways. It has been a mirror reflecting American values, a window into American life, and a canvas for American dreams.

However, the narrative of global cinema is changing. The

rise of film industries around the world and the increasing recognition of non-American films suggest a future where the cinematic stage is more diverse and inclusive. This evolution is a positive shift, one that allows for a richer exchange of ideas, stories, and cultural experiences.

The success of Hollywood and the emergence of global cinema are not mutually exclusive but are parts of the same story - the story of humanity's love for storytelling. It's a story that transcends borders, languages, and cultures. The future of cinema, therefore, lies not in one country or one style of filmmaking but in the collaborative and inclusive efforts of storytellers from around the world. As we look forward, we can anticipate a global cinema that continues to be shaped by the pioneering spirit of Hollywood, enriched by the diverse voices from across the globe. This is the future of storytelling - a tapestry woven from threads of different colors, textures, and patterns, reflecting the true diversity of the human experience.

SPACEX

In the annals of space exploration, SpaceX stands as a beacon of American ingenuity and ambition. Founded in 2002 by entrepreneur Elon Musk, SpaceX has not just entered the space race; it has revolutionized it. As the first private company to launch, orbit, and recover a spacecraft, SpaceX has shattered the paradigm of space travel being the sole domain of government entities. This is the story of a company that dared to dream big and in doing so, reignited the world's passion for the final frontier.

SpaceX's journey is more than a tale of technological prowess; it is a testament to the relentless pursuit of a vision that seemed impossible. It symbolizes a new era in space exploration, one where private enterprise leads the charge, pushing the boundaries of what is achievable and redefining humanity's place in the cosmos.

The Rise of SpaceX

In the early 2000s, the landscape of space exploration was dominated by governmental agencies with massive budgets. It was into this arena that SpaceX, a private company with bold ambitions and a fraction of the resources, entered. Founded by Elon Musk in 2002, SpaceX's mission was audacious: to reduce space transportation costs and enable the colonization of Mars. Musk, known for his successful ventures PayPal and Tesla, brought a vision to democratize space travel, making it accessible and sustainable. The early days of SpaceX were marked by high risks and skepticism. The idea of a private company not just participating in, but leading the charge in space exploration was met with doubts. However, Musk's determination and belief in his mission were unwavering.

SpaceX's first major goal was to develop a line of orbital

launch vehicles. The journey was fraught with challenges. The first three launches of the Falcon 1 rocket failed, bringing the company to the brink of bankruptcy. Despite these setbacks, the team persevered, driven by the belief in their mission. The fourth launch in 2008 was a success, marking a significant turning point. This achievement made SpaceX the first privately-funded company to send a liquid-fueled rocket to orbit, a milestone in space history. The success of Falcon 1 was just the beginning. It laid the groundwork for the development of the Falcon 9 and the Dragon spacecraft, further solidifying SpaceX's reputation as a game-changer in the industry. These early challenges and triumphs set the stage for a series of historic achievements that would not only redefine the company but also the future of space travel.

Major Milestones and Achievements

SpaceX, standing out as the only major private player in a field traditionally dominated by government-funded entities, has marked its journey with groundbreaking achievements. The company's innovative approach and risk-taking ethos have led to a series of unprecedented milestones in the world of space exploration. One of SpaceX's most significant accomplishments is the development of the Falcon 9 rocket. Its first successful launch in 2010 marked a new era for private space exploration. The Falcon 9 was not just any rocket; it was designed to be reusable, a feature that was revolutionary in the space industry. The successful landing of a Falcon 9 rocket's first stage in 2015 was a historic moment, drastically reducing the cost of accessing space.

Another major breakthrough came with the development of the Dragon spacecraft. In 2012, SpaceX's Dragon became the first privately-developed spacecraft to successfully dock with the International Space Station (ISS). This achievement was not only a technical feat but also symbolized a new collaborative era between private and government space programs. Furthermore, SpaceX has continually pushed the boundaries of what is possible in space travel. In 2018, the Falcon Heavy, the most powerful operational rocket in the world, successfully launched and sent a Tesla Roadster into space, showcasing both the rocket's capabilities and Musk's flair for dramatic innovation. Each of these milestones is not just a testament to SpaceX's engineering prowess but also highlights the company's role in pioneering a new model for space exploration. Where space was once the exclusive domain of national governments, SpaceX has shown that private companies can not only compete but also lead in the race to the stars.

Impact on American Space Exploration

SpaceX's emergence and success have had a profound impact on American space exploration. The company's achievements have not only propelled the United States back to the forefront of space endeavors but have also reshaped the landscape of space exploration as a whole.

One of the key aspects of SpaceX's impact is its partnership with NASA. In a significant shift from traditional models, NASA began contracting SpaceX for cargo and crew missions to the International Space Station (ISS). This collaboration is a prime example of a successful public-private partnership in space exploration. SpaceX's ability to provide cost-effective and reliable transport services has been a major asset to NASA, allowing the agency to allocate more resources to other scientific and exploratory missions. Moreover, SpaceX's innovations, particularly in rocket reusability, have sparked a new wave of enthusiasm and investment in the American space sector. The company's success has paved the way for other private entities to enter the space arena, fostering a competitive and dynamic industry.

The influence of SpaceX extends beyond logistics and technology; it has also played a role in shaping American space policy. The company's vision of Mars colonization and its tangible steps towards achieving it have reignited a sense of adventure and possibility in space exploration. This vision aligns with national interests in maintaining a leading role in space and has influenced policy discussions about future space exploration goals and methods. In essence, SpaceX has not only revolutionized space travel technologically but has also been instrumental in reinvigorating the American spirit of exploration, demonstrating that space is a frontier not just for governments but for all of humanity.

The Future of Space Exploration with SpaceX

As we look to the future, SpaceX's ambitious plans herald a new chapter in space exploration, one where America, the first nation to put a man on the Moon, now sets its sights on an even more daring goal: Mars. SpaceX's vision for the future is epitomized by the development of the Starship, a fully reusable spacecraft designed to carry humans to Mars and beyond. The Starship represents not just a leap in space technology but a rekindling of the pioneering spirit that first brought humanity to the Moon. It's a symbol of America's unwavering commitment to leading the way in space exploration.

The journey to Mars, as envisioned by SpaceX, is more than just a scientific endeavor; it's a step towards becoming a multiplanetary species. The Mars colonization plan, ambitious in its scope, aims to establish a sustainable human presence on the Red Planet. This endeavor, if successful, will be one of the most significant milestones in human history, demonstrating America's role as a trailblazer in pushing the boundaries of human exploration.

The potential impacts of SpaceX's Mars missions are profound. It's not just about the scientific discoveries and technological advancements; it's about inspiring a new generation of explorers and thinkers. The Mars missions could lead to advancements in areas like renewable energy, life support systems, and even social structures that could benefit life on Earth.

SpaceX's trajectory from pioneering private space exploration to leading the charge towards Mars encapsulates the American ethos of exploration and innovation. The company's journey symbolizes America's enduring spirit of reaching for the stars, literally and metaphorically, and maintaining its position at the forefront of space exploration.

The Journey to Mars

The story of SpaceX is more than a chronicle of a private company's foray into space; it's a testament to American ingenuity and the relentless pursuit of the extraordinary. From its audacious beginnings to its groundbreaking achievements, SpaceX has redefined what is possible in space exploration, embodying the spirit of American innovation and adventure.

SpaceX's journey reflects a broader narrative of American leadership in space. Just as America was the first to leave human footprints on the Moon, it now stands at the vanguard of the next great leap in space exploration – the journey to Mars. This endeavor, spearheaded by a private company, marks a significant shift in how space exploration is conducted and who participates in it. It symbolizes a future where space is not just a realm for governments but a frontier accessible to all of humanity.

The impact of SpaceX's achievements extends beyond space travel. It has reignited a global interest in space and science, inspiring people across the world to look up at the stars and dream of possibilities. The company's vision of Mars colonization stretches our imagination and encourages us to think about our future as a species in new and exciting ways.

SpaceX's story is a vivid illustration of the American spirit in action. It showcases the nation's ability to lead, innovate, and inspire in the quest to explore the unknown. As SpaceX continues to push the boundaries of space exploration, it carries with it the hopes, dreams, and aspirations of not just a nation but the entire human race, reminding us all that the sky is not the limit – it's just the beginning.

NEURALINK

Neuralink, a neurotechnology company founded by visionary entrepreneur Elon Musk in 2016 in San Francisco, is the newest American addition to the world. Its goal is to develop implantable brain–machine interfaces (BMIs) to help people with neurological conditions and eventually enhance human cognition.

In the heart of San Francisco, amidst the bustling streets and towering skyscrapers, lies the headquarters of a company that could redefine the future of humanity. Neuralink stands at the forefront of a new era in neurotechnology. This story is not just about a company or its revolutionary product; it's about an American invention that holds the potential to impact the world, akin to the great innovations that have consistently emerged from the United States. From the Wright brothers' first flight to the creation of the internet, American ingenuity has repeatedly reshaped how humans live, work, and communicate. Now, Neuralink aspires to join these ranks by bridging the gap between the human brain and the digital world. Its potential stretches far beyond the realms of science fiction, hinting at a future where the limitations of the human body and mind can be transcended. In a world where the extraordinary becomes the norm, Neuralink's journey is a testament to the relentless pursuit of progress that characterizes the American spirit. It's a story of ambition, challenges, and the relentless pursuit of a dream that could change the course of human history.

The Genesis of Neuralink

In the year 2016, Neuralink embarked on a journey that was as audacious as it was inspiring. Nestled in the innovative hub of San Francisco, a group of the world's brightest minds came together under the leadership of Elon Musk. Musk, already renowned for his role in revolutionizing space travel with SpaceX and altering the automotive industry with Tesla, was about to venture into uncharted territory. Neuralink was not just another tech startup; it was a beacon of futuristic aspirations, aiming to merge human consciousness with artificial intelligence.

Musk's vision was clear yet profoundly complex: to create a seamless interface between the human brain and computers. He foresaw a future where limitations of the human brain could be overcome, not only aiding those with neurological disorders but also enhancing human cognition. This vision was deeply rooted in the American ethos of pushing boundaries and creating innovations that could change the world.

The early days of Neuralink were filled with challenges. The team, comprising of neuroscientists, engineers, and other specialists, faced the monumental task of developing a device that could safely and effectively connect the human brain to a computer. They grappled with questions that seemed straight out of a science fiction novel. How could they create an interface delicate enough to interact with neurons? What materials could be used to prevent rejection by the human body? How could this technology be made accessible and ethical?

Despite these hurdles, Neuralink's team persevered, driven by the potential global impact of their work. They were not just creating a product; they were pioneering a technology that could redefine the human experience. The promise of Neuralink extended beyond the borders of the United States; it held the potential to address neurological diseases that affected millions worldwide, offering hope where there was none.

The development of Neuralink thus became a symbol of American innovation's positive influence on the world. It echoed the spirit of past inventions like the polio vaccine and the internet, which, though born in the United States, became global beacons of progress. Neuralink, still in its developmental stages, was poised to join this illustrious list, embodying the relentless American spirit of pushing the boundaries of what is possible.

The Technology Behind Neuralink

Imagine being able to control a computer or a smartphone just by thinking about it. This might sound like magic, but it's exactly what Neuralink is trying to achieve. At the heart of Neuralink's technology is something called a brain-machine interface (BMI). This is a tiny device that can be placed inside the brain, and it works like a translator, converting thoughts into actions on a computer or other devices.

To understand how Neuralink's BMI works is not easy. But we have to refresh our memory, just a couple decades ago it would not be easy to understand how we can talk over a phone and see each other from different parts of the world. Now, with Nearalink, think of the brain as a very busy office, with neurons (brain cells) as the office workers. These workers are constantly sending messages to each other, just like how neurons communicate through electrical signals. Neuralink's device acts like a super-smart office assistant, listening to these messages and translating them into a language that computers can understand.

Developing this technology wasn't easy. The team had to ensure that the device was small and safe enough to be placed in the brain without causing harm. They also had to make sure it was powerful enough to read the brain's complex signals. The device uses very thin threads, thinner than a human hair, which can read signals from neurons. These threads are connected to a small chip, which processes the signals and sends them to a computer.

The potential of Neuralink's technology is huge. For people with paralysis or other mobility issues, it could mean controlling wheelchairs or typing messages using only their thoughts. For others, it could eventually lead to new ways of interacting with technology, making life easier and more connected. Neuralink is more than just a high-tech project; it's a new chapter in the story of American innovation. Just like the invention of the light bulb or the internet, Neuralink has the potential to change the world in ways we can't even imagine yet. It's a testament to the power of human ingenuity and the endless possibilities that come with it.

Global Impact and Ethical Considerations

The potential global impact of Neuralink is as vast as the human imagination. At its core, Neuralink's technology could revolutionize how we approach some of the most challenging neurological disorders. Diseases like Parkinson's, epilepsy, and even depression, which affect millions worldwide, could be better managed or even alleviated. This isn't just a leap in medical science; it's a beacon of hope for those who have long felt forgotten by the rapid pace of technology.

Beyond medical applications, Neuralink promises to break down barriers in communication and learning. Imagine a world where language barriers are non-existent because thoughts can be directly translated into any language. Or consider the possibilities for education, where learning disabilities are no longer obstacles, as knowledge can be assimilated in new, more effective ways.

The technology could also be a significant equalizer in terms of access to technology. For people with physical disabilities, Neuralink could offer new avenues for interaction with the digital world, making technology more inclusive and accessible. However, with great power comes great responsibility. Neuralink's advancements raise important ethical questions. How do we ensure privacy and security when thoughts can potentially be read by machines? What are the implications for individual autonomy and identity? Neuralink is not just navigating technological challenges but is also setting precedents for the ethical use of this technology.

Internationally, Neuralink stands as a testament to the positive influence of American innovation. It embodies the spirit of pushing the boundaries for the greater good, a theme that resonates with the core values of numerous American breakthroughs that have benefitted humanity. In a world often divided by borders and differences, Neuralink serves as a reminder of what can be achieved when we focus on what unites us: our shared human experience and our collective pursuit of a better, more inclusive future.

The Future with Neuralink

Let's imagine the world in a few years, with Neuralink fully developed and integrated into our lives. It's like stepping into a future where the line between human and machine starts to blur, but in a good way.

Firstly, for people with disabilities, Neuralink could be life-changing. Imagine someone who can't move or speak being able to communicate and interact with the world just by thinking. It's like giving them a new voice, a way to express themselves and connect with others. This technology could open doors to opportunities that were once thought impossible.

In education, learning could become more personalized and efficient. Students struggling with certain subjects might be able to understand complex concepts more easily. It's like having a teacher who knows exactly how your brain works and tailors the lessons just for you. For healthcare, doctors might be able to diagnose and treat brain disorders more effectively. Conditions like Alzheimer's or brain injuries could be better understood, potentially slowing down or even reversing their effects.

But it's not just about solving problems. Neuralink could enhance the way we experience the world. Maybe you could learn a new language in days instead of months, or play a musical instrument without ever having practiced it. It's like unlocking a part of your brain that lets you learn and experience things in a whole new way. Of course, with all these advances, it's important to make sure that this technology is used responsibly and ethically. It should be accessible to everyone who needs it and used to improve lives, not control them.

In this future, Neuralink isn't just an American invention; it's a global innovation. It's a symbol of hope and progress, showing what humanity can achieve when we use technology to enhance our lives and work together for a better world. It's a future that's bright, inclusive, and full of possibilities, a true testament to the spirit of human ingenuity and the endless quest for improvement.

PART FOUR

AMERICA –
Immigrant's Angle

AMERICA - IMMIGRANT'S ANGLE

I'm not here to rediscover America, as Christopher Columbus already did in 1492. Instead, my intention is to remind all of us of the profound impact America has had on our lives, even though we tend to overlook it on a daily basis. I want to encourage us to take a moment and ponder how different our lives on this planet would be if America had never existed.

When I first arrived in America at the beginning of the 21st century, I was struck by the hustle and bustle of this land. It was teeming with people from all walks of life, and it boasted an expansive and breathtaking landscape. To me, America appeared as a place where dreams could come true, and possibilities seemed endless.

However, I didn't fully grasp at the time just how influential America was on a global scale. Over the years, I've come to realize that so much of what has shaped our world in the last century can be traced back to this very place. It's quite fascinating to think about how many significant developments, both big and small, have originated here in America and have impacted every single person on Earth. These are things that we often take for granted, yet they play a crucial role in making our daily lives easier and more comfortable.

The average human life spans less than 100 years, and it's astonishing to consider the incredible innovations, discoveries, advancements, and changes that have occurred in America over the past 100 to 150 years. These achievements are beyond our wildest imagination and have left an indelible mark on our world.

THE EPICENTER OF PLANET EARTH

America, the land I now call home, is more than just a country; it's the epicenter of our planet. Almost every facet of life on Earth in the last century has been influenced in one way or another by America. From technological marvels to cultural phenomena, the ripple effect of America's contributions can be felt across the globe.

Let's start with technology. America has been at the forefront of technological advancements for decades. Think about your smartphone – that incredible device that fits in your pocket and connects you to the world. It was born out of American innovation. Companies like Apple, Google, and Microsoft have shaped the way we communicate, work, and play. Moreover, the internet, the backbone of modern communication, has its roots in American research. The World Wide Web, created by Sir Tim Berners-Lee, revolutionized the way we access information and interact with each other. Social media platforms like Facebook and Twitter(X) have transformed the way we connect and share our lives.

America has also been a leader in the field of medicine. Groundbreaking discoveries and innovations in healthcare have improved the quality and length of our lives. Think about vaccines, which have saved countless lives worldwide. Many of the most essential vaccines, including those for polio, measles, and COVID-19, were developed in America. Medical technology has also made incredible strides in America. Advanced diagnostic tools, surgical techniques, and life-saving treatments are all products of American ingenuity. The development of organ transplantation, MRI machines, and minimally invasive surgeries has given hope to patients around the world.

America's influence extends beyond our planet's surface. The United States has played a central role in space exploration, starting with the moon landing in 1969. NASA, the American space agency, has been at the forefront of space missions, including the Mars rovers and the International Space Station. These missions have expanded our understanding of the universe and opened up possibilities for future space exploration. Beyond technology and science, American culture has permeated the globe. From Hollywood films to American music, fashion, and food, the influence of American culture is everywhere. Iconic American figures like Elvis Presley, Marilyn Monroe, and Michael Jackson have become global symbols of entertainment and artistry. American literature, from Mark Twain to F. Scott Fitzgerald, has left an indelible mark on the literary world. The American dream has been a source of inspiration for people worldwide.

America's economic impact is undeniable. It is home to some of the world's largest and most influential companies, including Amazon, Apple, and Microsoft. These corporations not only drive innovation but also shape the global economy. The New York Stock Exchange and NASDAQ are among the world's largest stock exchanges, playing a pivotal role in global finance. America's political system has also had a significant impact on the world. The United States' democratic principles have served as a model for many nations. The idea of individual rights and freedoms, enshrined in the U.S. Constitution, has influenced the development of democracies around the world. America has a long history of humanitarian efforts. Whether it's providing aid during natural disasters, offering refuge to those fleeing persecution, or leading global initiatives to combat disease and poverty, the United States has consistently played a vital role in making the world a better place. In recent years, America has also taken strides in addressing environmental challenges. The country has been a driving force behind global efforts to combat climate change. Initiatives to reduce greenhouse gas emissions and transition to renewable energy sources have the potential to positively impact the entire planet.

If we take a moment to imagine a world where America never existed, it's a challenging exercise, but it helps us appreciate the immense contributions this nation has made to our lives. Without America, the world might not have witnessed the technological marvels that have become integral to our existence. Your smartphone might never have been invented, and you'd be relying on older, less efficient forms of communication and information retrieval.

Medical progress would also be significantly hindered. Vaccines and life-saving treatments might have been delayed or never discovered, leading to higher mortality rates from preventable diseases. The absence of American contributions to healthcare would leave us with fewer tools to fight illnesses. **Space exploration** would look very different, with fewer missions and less understanding of the cosmos. The inspiring achievements of NASA, from the moon landing to exploring distant planets, would be absent from our history. The influence of **American culture** would be sorely missed. Hollywood blockbusters, iconic music, and the global reach of American fashion would be replaced with a different cultural landscape. American literature, which has enriched our lives with timeless classics, would never have been written. Economically, the absence of America would create a void in the **global economy**. Major corporations that drive innovation and create jobs worldwide would cease to exist, potentially leading to economic stagnation and fewer opportunities for people across the globe.

The **political influence** of America, with its democratic principles and advocacy for human rights, would leave a void in the global stage. The world might be a very different place, with potentially fewer democracies and less emphasis on individual freedoms. **Humanitarian efforts and environmental** initiatives, often spearheaded by American leadership, would face significant setbacks. The world might struggle to respond to crises and address pressing environmental challenges effectively.

In essence, the world without America would be a vastly different and less advanced place. The contributions and innovations that have shaped our modern lives would be conspicuously absent, leaving us with a world that is less interconnected, less prosperous, and less capable of addressing global challenges. America's ability to innovate and adapt has been a driving force behind its global influence. The nation's commitment to research and development, entrepreneurship, and a culture that encourages risk-taking have all played a significant role in shaping the world we live in today.

THE SPIRIT OF INNOVATION

Innovation is at the heart of America's success story. It's a nation that thrives on pushing boundaries, challenging the status quo, and seeking new solutions to old problems. The spirit of innovation is deeply ingrained in American society and has been a driving force behind many of the world-changing developments we often take for granted. One key element of this spirit is the culture of entrepreneurship. America has long been a haven for entrepreneurs and innovators from around the world. It's a place where individuals with big ideas and dreams can find the resources, support, and opportunities to turn their visions into reality. Silicon Valley, located in California, is a prime example of this entrepreneurial culture, known as the global hub for technology innovation and startup success.

American entrepreneurs have been responsible for creating some of the most influential companies in the world. Names like Nikola Tesla, Thomas Edison, Henry Ford, Steve Jobs, and Elon Musk are synonymous with innovation and have left an indelible mark on various industries. These individuals didn't just invent products; they revolutionized entire industries, from electricity to automobiles and personal computers to space exploration. Let's delve deeper into some specific areas where American innovation has had a profound impact.

Technology and Communication

As mentioned earlier, the world of technology and communication has been dramatically shaped by American innovation. The development of the internet, a project initially funded by the U.S. government, has transformed the way we access information and connect with one another. Email, social media, and online shopping have all become integral parts of our daily lives, thanks to American ingenuity. Moreover, the rise of the smartphone, with companies like Apple and Google leading the way, has changed how we work, play, and communicate. These pocket-sized devices serve as our communication hubs, entertainment centers, and productivity tools. The mobile app ecosystem, another American creation, has opened up new possibilities and industries, from ride-sharing to food delivery.

Healthcare and Medicine

American innovation has been a driving force in healthcare and medicine. The development of vaccines, which prevent deadly diseases like polio, measles, and COVID-19, has saved countless lives worldwide. American pharmaceutical companies have pioneered the research and development of these vaccines, making them accessible to people across the globe. In addition to vaccines, medical technology has made significant advancements. From the invention of the MRI machine to the development of minimally invasive surgical techniques, American innovation has improved patient outcomes and reduced the invasiveness of medical procedures. Organ transplantation, a medical miracle that gives people a second chance at life, was made possible through American research and innovation.

Space Exploration

America's role in space exploration cannot be overstated. The historic moon landing in 1969, when astronauts Neil Armstrong and Buzz Aldrin set foot on the lunar surface, was a momentous achievement. NASA, the American space agency, has continued to explore the cosmos, sending missions to Mars, Jupiter, and beyond. These missions expand our understanding of the universe and inspire future generations of scientists and explorers. Furthermore, private American companies like SpaceX, founded by Elon Musk, are pushing the boundaries of space exploration. They are working on making space travel more accessible to ordinary people and envisioning a future where humans can inhabit other planets. Such ambitious projects have the potential to reshape our understanding of space and humanity's place in it.

Cultural Influence

American culture has a global reach like no other. From Hollywood blockbusters to iconic music, fashion trends, and culinary innovations, American culture permeates societies worldwide. The influence of American pop culture extends beyond entertainment; it shapes the way people dress, the music they listen to, and the food they eat. For instance,

American fast food chains, such as McDonald's and Starbucks, have become ubiquitous in cities around the world. The concept of fast food itself, with its quick and convenient dining options, has been exported globally, changing the way people eat and dine out. Moreover, American literature has produced timeless classics that have resonated with readers worldwide. Works like "To Kill a Mockingbird" by Harper Lee and "The Great Gatsby" by F. Scott Fitzgerald have left a lasting impact on literature and continue to be studied and celebrated.

Economic Powerhouse

America's economic influence extends far and wide. It is home to some of the world's largest and most influential corporations, including tech giants like Amazon, Apple, and Microsoft. These companies not only drive innovation but also shape the global economy. They create jobs, generate wealth, and contribute to economic growth not only in the United States but also in countries with whom they do business. The New York Stock Exchange (NYSE) and NASDAQ, both American institutions, are among the world's largest stock exchanges. They provide a platform for companies to raise capital and for investors to buy and sell shares. These exchanges play a pivotal role in global finance, influencing stock markets and investment strategies worldwide.

Political Influence

America's political system, rooted in democratic principles and individual rights, has served as a model for many nations. The U.S. Constitution, a foundational document in American history, has inspired the development of democratic governments around the world. The idea of a government accountable to its citizens and the protection of individual freedoms has become a global aspiration. The United States has also been a strong advocate for human rights and democracy on the international stage. It has supported efforts to promote democratic governance, protect civil liberties, and address humanitarian crises. American leadership in organizations like the United Nations has played a crucial role in shaping global policies and responses to pressing issues.

Humanitarian Efforts

America has a long history of humanitarian efforts, both within its borders and abroad. When natural disasters strike, American organizations and agencies are often among the first to provide aid and relief. Whether it's responding to hurricanes, earthquakes, or wildfires, the United States consistently extends a helping hand to those in need. Furthermore, America has been a refuge for individuals fleeing persecution and seeking a better life. The nation's history includes waves of immigrants who have contributed to its cultural diversity and economic vitality. American generosity and compassion have also extended to refugees and asylum-seekers, providing them with opportunities for a fresh start. On the global stage, America has taken a lead role in addressing humanitarian crises and providing aid to countries facing challenges such as famine, disease outbreaks, and conflict. American assistance has saved lives, alleviated suffering, and supported development in some of the world's most vulnerable regions.

Environmental Initiatives

In recent years, America has taken significant steps to

address environmental challenges. The country's commitment to environmental sustainability and efforts to combat climate change have global implications. Initiatives to reduce greenhouse gas emissions, transition to renewable energy sources, and protect natural habitats have the potential to positively impact the entire planet. American innovation in renewable energy technologies, such as solar and wind power, has accelerated the shift toward a more sustainable future. Electric vehicles (EVs) have gained traction as an eco-friendly alternative to traditional gasoline-powered cars, with American companies like Tesla leading the way. These developments align with global efforts to reduce carbon emissions and mitigate the effects of climate change.

AN ALTERNATE WORLD WITHOUT AMERICA

Now that we've explored the incredible contributions of American innovation across various fields, let's return to the thought experiment of imagining a world without America. This exercise highlights the profound impact that American innovation has had on our lives and the world as a whole.

Transportation and Infrastructure

In a world without America, transportation and infrastructure would look significantly different. American innovations like the assembly line, pioneered by Henry Ford, revolutionized the automobile industry. This innovation made cars more affordable and accessible to the masses, leading to the global prevalence of personal automobiles. The absence of American automotive companies like Ford, General Motors, and Chrysler would leave a void in the market. Alternatives from other parts of the world might exist, but the pace of innovation and affordability that Americans brought to the industry would be sorely missed.

Additionally, the extensive network of highways and infrastructure development in America has served as a model for transportation systems worldwide. The Interstate Highway System, initiated in the 1950s, transformed how people and goods move across the country. Its impact on logistics, trade, and economic development cannot be overstated.

Aviation and Aerospace

American contributions to aviation and aerospace have been monumental. The Wright brothers, Orville and Wilbur, achieved the first powered, controlled flight in 1903. This historic event marked the beginning of modern aviation and set the stage for the development of commercial aviation, which connects people and places around the world.

Without American aviation pioneers and companies like Boeing, Airbus, and Lockheed Martin, the progress in aerospace technology and air travel would likely have been significantly delayed. Commercial air travel, a cornerstone of global connectivity, might not be as advanced or widespread as it is today.

Energy Production and Distribution

America has played a pivotal role in the energy sector, from the discovery of oil in Texas to the development of advanced energy technologies. The American oil industry, which emerged in the late 19th century, transformed the world's energy landscape. The availability of abundant and affordable energy sources has been essential for economic growth and development. In a world without America, the global energy picture would be different. The development of alternative energy sources, such as natural gas and renewable energy, might have been delayed. The transition to cleaner and more sustainable energy options could face significant hurdles.

Agriculture and Food Production

American innovations in agriculture have had a profound impact on global food production. Innovations like the cotton gin, developed by Eli Whitney, revolutionized the textile industry. The widespread adoption of mechanized farming equipment, such as the combine harvester, increased agricultural productivity and food availability. The Green Revolution, driven by American scientists and researchers, introduced high-yield crop varieties and improved agricultural practices. This initiative has been instrumental in alleviating food shortages and hunger in many parts of the world. Without American agricultural innovations, food production and distribution would likely be less efficient, and food security could be a more significant global challenge.

Education and Research

American universities and research institutions are renowned for their contributions to knowledge and innovation. They attract talent from around the world and conduct groundbreaking research in various fields. American universities have been at the forefront of technological advancements, medical discoveries, and scientific breakthroughs. In a world without America, the global landscape of education and research would be different. The absence of American universities and research institutions could slow down progress in various disciplines. Collaborative international research efforts would face limitations, potentially hindering the development of new technologies and solutions to global problems.

Global Governance and Diplomacy

America has played a central role in global governance and diplomacy. It is a founding member of the United Nations and a permanent member of the United Nations Security Council. American leadership has been instrumental in addressing international conflicts, humanitarian crises, and peacekeeping efforts. In a world without America, the balance of power in international relations would shift. The absence of American diplomatic initiatives and peacekeeping missions could lead to different outcomes in global conflicts. The international community might need to adapt to a world where American influence is less pronounced.

Art and Entertainment

American contributions to the world of art and entertainment are undeniable. Hollywood, the epicenter of the global film industry, produces a vast majority of the world's most popular movies and television shows. American actors, directors, and writers have left an indelible mark on the art of storytelling through film. American music, from jazz to rock 'n' roll to hip-hop to pop, has influenced musical genres worldwide. Artists like Elvis Presley and Michael Jackson are cultural icons celebrated on every continent. In a world without America, the cultural landscape would be vastly different. The absence of Hollywood would mean fewer blockbuster movies and a different cinematic experience.

Music genres and styles that originated in America might not have reached the same level of global recognition.

The Role of Global Partners

It's essential to recognize that American innovation has often been a collaborative effort, involving individuals and ideas from around the world. Many innovations attributed to America have roots in international contributions. Scientists, engineers, and entrepreneurs from diverse backgrounds have contributed to American progress. In a world without America, other nations might have taken on greater roles in innovation and leadership. Different countries and regions might have emerged as hubs for technological advancement, scientific research, and cultural influence.

However, the pace and scale of innovation could be different without the unique blend of factors that have characterized American society, including a robust entrepreneurial culture, access to resources, and a commitment to research and development. The impact of America on the world cannot be overstated. American innovation has shaped the way we live, work, communicate, and interact with the world around us. From technology and healthcare to space exploration and cultural influence, America's contributions have been at the forefront of global progress.

The thought experiment of imagining a world without America underscores the incredible achievements and innovations that we often take for granted. It reminds us of the significant role America has played in advancing humanity and addressing global challenges. As we reflect on the past and look to the future, it's crucial to recognize and appreciate the contributions of America while also acknowledging the collaborative efforts of individuals and nations worldwide. Together, we continue to push the boundaries of knowledge, innovation, and progress, shaping a brighter and more interconnected world for generations to come.

GLOBAL PARTNERSHIP

Throughout history, collaboration and partnerships between America and other nations have led to groundbreaking innovations and advancements. International cooperation in science, technology, and various fields has been pivotal in addressing global challenges.

American scientists and researchers have frequently collaborated with their counterparts from around the world. Scientific discoveries and breakthroughs are often the result of collective efforts. The international exchange of knowledge and expertise has accelerated progress in various fields. For example, the discovery of the Higgs boson, a fundamental particle in particle physics, involved contributions from scientists representing numerous countries. Large-scale scientific experiments and projects, such as the Large Hadron Collider (LHC) in Europe, have brought together scientists from diverse backgrounds to explore the frontiers of knowledge.

Space exploration has been a symbol of international cooperation. While America has been a leader in space missions, it has also worked closely with other countries and space agencies. The International Space Station (ISS) stands as a testament to global collaboration in space exploration. The ISS is a joint project involving NASA (United States), Roscosmos (Russia), ESA (European Space Agency), JAXA (Japan Aerospace Exploration Agency), and CSA (Canadian Space Agency). Astronauts from various nations live and work together on the ISS, conducting scientific research that benefits humanity as a whole.

Global health challenges, such as the fight against infectious diseases and pandemics, demand international cooperation. American organizations and institutions have partnered with global health agencies, governments, and NGOs to combat diseases like HIV/AIDS, malaria, and Ebola. For instance, the President's Emergency Plan for AIDS Relief (PEPFAR), initiated by President George W. Bush, has provided support and resources to countries heavily affected by HIV/AIDS. This program, along with international partners, has made significant strides in reducing the impact of the disease.

Environmental conservation is another area where international cooperation is critical. America has participated in global efforts to address climate change and protect biodiversity. The Paris Agreement, an international treaty to limit global warming, is an example of collective action. While American leadership in environmental initiatives is significant, the commitment of countries worldwide is essential to combatting climate change and preserving the planet's natural resources.

American universities have long been a magnet for international students seeking higher education. These students contribute to diverse perspectives on campus and enrich the academic environment. Many of them return to their home countries with new skills and ideas, benefiting their societies. Moreover, American institutions offer scholarships and exchange programs that allow students from various backgrounds to study in the United States. This cultural exchange fosters goodwill and understanding between nations, strengthening international ties. Education and cultural exchange programs have played a vital role in fostering understanding and collaboration between nations. American universities and institutions have welcomed students and scholars from around the world, facilitating the exchange of ideas and knowledge.

American cultural diplomacy efforts, including programs like the Fulbright Program and the Peace Corps, promote cross-cultural understanding and collaboration. These initiatives send American scholars, artists, and volunteers abroad to share their expertise and learn from other cultures. Cultural exchange not only deepens mutual understanding but also contributes to soft power—the ability to influence others through attraction and appeal rather than coercion. American music, films, literature, and art have transcended borders, becoming a source of inspiration and cultural exchange worldwide.

As we consider the impact of American innovation and collaboration on the world, it's essential to recognize that the global landscape is continually evolving. New challenges and opportunities emerge, requiring nations to adapt and cooperate in different ways. Emerging technologies, such as artificial intelligence (AI), biotechnology, and quantum computing, are reshaping industries and societies. American companies and research institutions are at the forefront of these advancements, but the global nature of innovation means that expertise and contributions come from around the world. Collaborative research and development efforts in emerging technologies involve scientists and engineers from diverse backgrounds. International standards and regulations are essential to ensure the responsible and ethical use of these technologies on a global scale.

The COVID-19 pandemic underscored the importance of global health security. The rapid spread of the virus across borders highlighted the interconnectedness of nations and the need for coordinated responses. International organizations like the World Health Organization (WHO) played a crucial role in monitoring the pandemic and facilitating information sharing among countries. Collaborative efforts in vaccine development, production, and distribution have been essential in the fight against COVID-19.

Environmental challenges, particularly climate change, remain a pressing global concern. The United States, under various administrations, has taken steps to rejoin international climate agreements, such as the Paris Agreement, signaling a commitment to addressing climate-related issues. Global efforts to reduce carbon emissions, transition to renewable energy sources, and protect ecosystems are vital to mitigating the impacts of climate change. Collaborative research and technology sharing are key components of finding sustainable solutions.

A SHARED FUTURE

The impact of America on the world extends beyond its borders. American innovation, collaboration, and contributions to various fields have shaped the global landscape in profound ways. While America has played a pivotal role in advancing knowledge, technology, and culture, it is essential to recognize that the world's progress is a collective endeavor.

As we move forward, facing new challenges and opportunities, the importance of international cooperation and collaboration cannot be overstated. A shared future, where nations work together to address global issues and harness the power of innovation, offers the best path toward a brighter and more interconnected world for generations to come. By acknowledging the contributions of America and embracing a spirit of global cooperation, we can build a future that transcends borders and fosters a better understanding of our shared humanity. In doing so, we honor the legacy of innovation and progress that has defined America's role in the world and look forward to a world of continued collaboration and shared achievements.

PART FIVE

AMERICAN ICONS

AMERICAN ICONS

America, a land of endless possibilities, has been a nurturing ground for some of the most influential figures in modern history. Their diverse backgrounds and unique talents have not only shaped America but also left an indelible mark on the world.

The United States of America stands as a beacon of hope and a land of opportunity, a place where dreams are not just born, but are also given the chance to flourish. This nation, founded on principles of freedom, innovation, and diversity, has consistently provided a fertile ground for some of the most brilliant minds and influential personalities in history. These individuals, originating from various backgrounds, ethnicities, and beliefs, have not only shaped the American ethos but have also left an indelible mark on the world. Their stories, varied and inspiring, serve as powerful testaments to the notion that in America, the potential to achieve greatness knows no bounds. Here's a look at these remarkable individuals.

1. **Albert Einstein**, a theoretical physicist, who developed the theory of relativity, one of the two pillars of modern physics, is a paragon of scientific ingenuity. His work not only transformed our understanding of the universe but also underscored the United States' role as a sanctuary for scientific thought and discovery.

2. **Nikola Tesla**, an inventor and engineer, known for his developments in the field of electromagnetism in the late 19th and early 20th centuries, epitomizes the spirit of American innovation. Tesla's work laid the foundation for modern alternating current electric power systems, deeply influencing the technological landscape of the modern era.

3. **Thomas Edison**, an inventor and businessman, developed many devices in fields such as electric power generation, mass communication, sound recording, and motion pictures. His inventions, which include the phonograph and the long-lasting, practical electric light bulb, have had a lasting impact on the world, highlighting the fertile environment for innovation that America fosters.

4. **Elon Musk**, a business magnate, industrial designer, and engineer, is a contemporary example of the American dream in action. As the founder of SpaceX, CEO of Tesla, Inc., and co-founder of Neuralink and OpenAI, Musk has been a pivotal figure in the advancement of electric vehicles, sustainable energy, and space technology, demonstrating America's leading role in the forefront of technological innovation.

5. **Steve Jobs**, co-founder of Apple Inc., was an

innovator in technology, transforming the computer, music, and mobile industries. His vision in creating products like the iPhone, iPad, and Mac revolutionized our way of living and interacting, showcasing America's unique ability to turn dreams into tangible, world-changing technologies.

6. **Bill Gates**, best known as the co-founder of Microsoft Corporation, played a significant role in the digital revolution. Today, Gates is also known for his philanthropic work in global health, education, and climate change. His journey from a passionate technologist to a philanthropist embodies the American ethos of innovation, hard work, and giving back to society.

7. **Mark Zuckerberg**, as the co-founder of Facebook, redefined global communication. His work in creating the most extensive social network connects billions, exemplifying America's role in the digital age as a creator of platforms that bring people together from all corners of the globe.

8. **Jeff Bezos**, the founder of Amazon.com, transformed the e-commerce industry. His vision and innovation in online shopping and cloud computing have not only changed how we buy goods but also how businesses operate, showcasing the expansive potential of American entrepreneurial spirit.

9. **Ernest Hemingway**, a Nobel Prize-winning novelist, known for his terse and straightforward writing style, has influenced generations of writers and readers alike. His works, which reflect his adventurous life and

unique view of the world, underscore America's contribution to literature and the arts.

10. **Martin Luther King Jr.**, an American Baptist minister and activist, became the most visible spokesperson and leader in the civil rights movement from 1955 until his assassination in 1968. King's advocacy for civil rights, based on principles of nonviolence and civil disobedience, changed the course of American history and inspired movements for civil rights and freedom across the globe.

11. **John F. Kennedy**, the 35th president of the United States, left a significant mark on American politics. His leadership during the Cuban Missile Crisis, his vision for space exploration, and his promotion of civil rights legislation were pivotal in shaping the nation's trajectory and reflect America's role as a leader on the global stage.

12. **Barack Obama**, the first African American president of the United States, broke barriers and embodied the idea that in America, anything is possible. His presidency, marked by significant reforms in healthcare and an emphasis on inclusivity and diversity, is a testament to the evolving American dream.

13. **Donald Trump**, outsider in politics, also showed that anything is possible. A businessman and television personality before becoming the 45th president, brought an unconventional approach to the office. His presidency, marked by its distinctive style and policies, underscores the diverse nature of American democracy and the wide array of perspectives that shape it.

14. **Muhammad Ali**, an American professional boxer, activist, and philanthropist, was known as one of the most significant and celebrated sports figures of the 20th century. His achievements in the boxing ring and his activism for racial equality and social justice have made Ali not just a sports icon, but a symbol of the American struggle for civil rights and global humanitarian causes.

15. **Walt Disney**, an entrepreneur, animator, writer, voice actor, and film producer, transformed entertainment with his innovative ideas and visionary creative spirit. Disney's legacy in animation, theme parks, and family entertainment highlights America's unique ability to craft stories and experiences that resonate globally.

16. **Bruce Lee**, a martial artist, actor, director, and philosopher, broke down barriers in Hollywood for actors of Asian descent and popularized martial arts worldwide. Lee's influence extends beyond film and martial arts, symbolizing America's melting pot culture and the impact one individual can have on integrating diverse heritages.

17. **Oprah Winfrey**, a talk show host, television producer, actress, author, and philanthropist, is an example of how determination and talent can lead to extraordinary influence and success. Winfrey's journey from poverty to becoming North America's first black multi-billionaire is a narrative of American resilience and possibility.

18. **Elvis Presley**, known as the "King of Rock and Roll,"

his dynamic voice and energetic performance style are emblematic of American music's power and global appeal. Presley's cultural impact extended beyond music, influencing fashion, attitudes, and norms, reflecting America's role in shaping global pop culture.

19. **Michael Jackson**, an artist who earned the title "King of Pop," his contributions to music, dance, and fashion, along with his publicized personal life, made him a global figure in popular culture. Jackson's legacy in entertainment showcases America's ability to produce artists who transcend cultural, racial, and national boundaries.

20. **Bob Dylan**, a singer-songwriter and visual artist, is regarded as one of the greatest songwriters of all time. His work, which spans over six decades, is marked by its philosophical, social, and political themes, reflecting the American spirit of questioning and redefining the status quo.

21. **Elizabeth Taylor**, a global icon, epitomized Hollywood glamour with a career spanning over six decades. Renowned for her striking beauty and violet eyes, Taylor's performances in classics like "Cleopatra" and "Who's Afraid of Virginia Woolf?" earned her two Academy Awards. Beyond her cinematic success, she was a pioneering AIDS activist, leveraging her fame for philanthropy. Taylor's enduring legacy as a symbol of talent and humanitarianism cements her status as not just an American treasure, but a global legend.

22. **Madonna**, a singer, songwriter, and actress, known for her continuous reinvention as a music artist and her

versatility in music production, songwriting, and visual presentation. She is often referred to as the "Queen of Pop" and is noted for her cultural impact on the world stage, particularly in shaping modern music and female empowerment.

23. **Marilyn Monroe**, an actress, model, and singer, who became a leading sex symbol of the 1950s and early 1960s. Her films grossed over $200 million, a testament to her talent and appeal. Monroe remains an iconic figure in popular culture, representing America's influence in the realms of film, fashion, and celebrity culture.

These individuals, each with their unique contributions, underscore America's capacity to foster talent and innovation, regardless of one's background. Their stories are testaments to the idea that in America, the potential to achieve greatness and contribute to both national and global progress is boundless. This notion not only defines the American experience but also serves as an inspiration for people everywhere, proving that with determination and opportunity, the sky's the limit.

These luminaries, through their relentless pursuit of excellence and their unwavering belief in their dreams, encapsulate the essence of the American spirit. Their journeys are diverse, yet they share a common thread – a relentless pursuit of passion in a land that encourages, nurtures, and celebrates such endeavors.

America stands unique in its promise and its delivery of opportunities for those who dare to dream. It is a nation where the confluence of freedom, diversity, and the pursuit of happiness creates an environment ripe for individual success and global influence. The stories of these 23 Americans exemplify this extraordinary American ethos. Their legacies remind us that in America, talent coupled with determination can break barriers, redefine industries, and shape the world. This nation continues to be a place where dreams are not just welcomed but are given the chance to soar to their highest potential, making America and the world a better place.

PART SIX

SOCIAL HARDWARE

ARTIFICAL INTELLIGENCE (AI)

Artificial Intelligence (AI) is something many people around the world don't fully understand or feel comfortable with. Some have never even heard of it, while others have heard about it but are worried, doubtful, or scared of it. Most people simply don't know much about AI, and only a small group of individuals have a positive view of AI and its future in our world. This fear and uncertainty surrounding AI is not unusual, as history has shown that people often react this way to new things that originate from the United States.

What is Artificial Intelligence?

To start, let's explain what Artificial Intelligence actually is. At its core, AI refers to the ability of machines and computers to perform tasks that typically require human intelligence. These tasks include things like problem-solving, learning from experience, recognizing patterns, and making decisions. AI systems use advanced algorithms and large amounts of data to

243

perform these tasks efficiently.

AI can be divided into two main categories: narrow or weak AI and general or strong AI. Narrow AI is designed to perform a specific task, like voice recognition or image classification. It excels at the task it was created for but cannot perform other tasks outside its area of expertise. General AI, on the other hand, possesses human-like intelligence and can adapt to various tasks, similar to how a human can switch from driving a car to cooking a meal. General AI is still largely theoretical and has not been fully realized.

AI is already a part of our daily lives in various ways, even if we might not always be aware of it. Voice-activated virtual assistants like Siri, Google Assistant, and Alexa use AI to understand and respond to our voice commands. When you shop online, watch videos on platforms like Netflix, or listen to music on Spotify, AI algorithms analyze your past behavior to suggest products, movies, or songs that you might like. Self-driving cars use AI to navigate and make decisions on the road, aiming to reduce accidents and improve transportation efficiency. AI is used in medical diagnostics, helping doctors detect diseases from medical images like X-rays and MRIs more accurately. Translation apps like Google Translate use AI to translate text and speech from one language to another.

These examples demonstrate that AI has the potential to make our lives easier and more convenient. It can also improve efficiency in various industries, from healthcare to transportation. However, as with any transformative technology, there are concerns and reservations among the public.

The Historical Context of Fear

It's not the first time that people have been fearful of a new technology or concept, especially when it originates from the United States. Throughout history, we have seen similar patterns of fear and resistance to change.

When **electricity** was first introduced in the late 19th century, it was met with skepticism and fear. People were apprehensive about the idea of invisible energy flowing through wires and into their homes. They worried about safety and the potential for harm. However, over time, electricity transformed society, bringing light, power, and countless technological advancements.

The **internet**, which has become an integral part of our lives today, was met with suspicion when it emerged in the 20th century. People were concerned about privacy, online scams, and the potential for misinformation. Yet, the internet has revolutionized communication, information sharing, and business.

The invention of **automobiles** faced resistance as well. People were initially skeptical of this new mode of transportation, fearing accidents and the impact on traditional industries like horse breeding and blacksmithing. Despite the initial hesitation, cars have transformed the way we travel and shaped modern society.

When **airplanes** were first introduced, many questioned the safety of flying and were fearful of the unknown. Today, air travel is one of the safest and most efficient modes of transportation, connecting people across the globe.

Even life-saving medical advancements like **vaccines** faced resistance and skepticism. When vaccines were first developed to combat diseases, there were concerns about their safety and effectiveness. However, vaccines have played a crucial role in eradicating or controlling many deadly diseases.

These historical examples illustrate that initial fear and resistance to new technologies are not uncommon. Over time, as people become more familiar with these innovations and their benefits become evident, the fear subsides, and society adapts. The same process is likely to occur with Artificial Intelligence.

Birth of Artificial Intelligence

The story of artificial intelligence (AI) begins not in a lab or a workshop, but in the realm of imagination. It was the mid-20th century when American computer scientist and cognitive psychologist Allen Newell, along with his colleague Herbert A. Simon, made significant contributions to the field of AI. They developed the "Logic Theorist," often considered the first AI program, at the RAND Corporation in Santa Monica, California. In 1956, John McCarthy, an American computer scientist, organized the Dartmouth Conference, where the term "Artificial Intelligence" was first coined. This event, held at Dartmouth College in New Hampshire, marked the formal birth of AI as a field of research. It brought together researchers passionate about the concept of a machine that could mimic the human mind.

Fast forward to the late 20th century, Silicon Valley in California emerged as the epicenter of technological innovation, and AI was no exception. Stanford University became a significant hub for AI research, particularly under the guidance of John McCarthy, who continued to be a pioneering figure in the field. During the 1980s and 1990s, the rise of the internet and the digital age provided new impetus for AI research. American companies like IBM made headlines with their AI developments. In 1997, IBM's Deep Blue, a chess-playing computer, defeated world champion Garry Kasparov, showcasing the potential of AI in problem-solving and strategic thinking.

The turn of the century marked a new era in AI development. American universities and tech companies were at the forefront of this revolution. In 2012, a team led by Geoffrey Hinton, a British-American computer scientist and psychologist, made a breakthrough in deep learning at the University of Toronto, significantly impacting the AI field. American tech giants like Google, Apple, and Microsoft invested heavily in AI research and development. Google's acquisition of DeepMind in 2014, was a landmark moment in AI history.

As the 21st century progressed, the focus in AI shifted towards machine learning and big data, fields where American researchers and companies led significant advancements. In 2016, an important milestone was achieved when Google's DeepMind AI, AlphaGo, beat Lee Sedol, a world champion in Go, demonstrating the power of machine learning in complex problem-solving. This era also saw the rise of American personalities like Andrew Ng, a Chinese-American computer scientist. Co-founder of Google Brain, Ng played a crucial role in popularizing deep learning. His work at Stanford University, particularly in online education, helped democratize AI education, making it accessible to a broader audience.

AI in Everyday Life

AI began to permeate everyday life in America and globally. American companies like Tesla, led by Elon Musk, revolutionized the automotive industry with AI-powered self-driving cars. Meanwhile, AI in healthcare saw significant advancements, with companies like IBM's Watson Health using AI to assist in diagnostics and treatment planning. The influence of AI extended to entertainment and social media, with American companies like Netflix and Facebook utilizing AI algorithms for personalized recommendations and content curation. The impact of AI on everyday life was profound, reshaping how people interact with technology and each other.

As AI became more integrated into society, questions about ethics and responsibility emerged. American scholars and tech leaders, such as Fei-Fei Li, a Chinese-American computer scientist, and Timnit Gebru, an Ethiopian-American researcher, contributed significantly to the discourse on ethical AI. Their work highlighted the need for AI to be fair, unbiased, and transparent. The future of AI, particularly in America, is geared towards responsible and ethical development. Initiatives like the AI Now Institute at New York University focus on the social implications of AI, ensuring that as technology advances, it aligns with human values and societal needs.

As the 21st century progressed, AI's integration into various industries showcased American innovation and leadership. In finance, American firms like JPMorgan Chase and Goldman Sachs adopted AI for risk management and algorithmic trading, revolutionizing the way financial markets operate. In agriculture, AI technologies developed by American companies like John Deere facilitated precision farming, using data and AI to enhance crop yield and sustainability. This integration signified a leap in traditional sectors, merging them with cutting-edge technology.

American space exploration and defense sectors have also been significantly influenced by AI. NASA's use of AI in missions, such as the Mars Rovers, demonstrated the potential of AI in navigating and analyzing extraterrestrial environments. In defense, the Pentagon's investment in AI for national security and defense strategies underscored the strategic importance of AI in modern warfare and security. In this journey of AI evolution, several American figures have been pivotal. Elon Musk, while known for his work in electric vehicles and space, has also been a vocal advocate for the responsible use of AI. OpenAI focuses on ensuring that AI benefits all of humanity. Regina Dugan, an American businesswoman and technology developer, led advanced technology projects at Google and Facebook, pushing the boundaries of AI applications. Her work exemplifies the American spirit of innovation and risk-taking in the field of AI.

As AI continues to evolve, the American dream of innovation and progress remains central to its development. The United States, home to some of the world's leading tech companies and research institutions, continues to be at the forefront of AI research and application. Looking ahead, the focus is on developing AI that is not only technologically advanced but also socially responsible and ethically sound. The challenge for American innovators and policymakers is to balance rapid technological advancement with the broader implications for society and humanity.

The Future of AI

The future of AI is full of promise and potential benefits. AI has the capacity to address some of the most pressing global challenges, from climate change and healthcare to poverty alleviation and education.

AI can assist doctors in diagnosing diseases more accurately and quickly, leading to better patient outcomes. It can also help in drug discovery and personalized medicine. AI can analyze vast amounts of environmental data to monitor and mitigate climate change, protect endangered species, and manage natural resources more sustainably. Personalized learning platforms powered by AI can tailor educational content to individual students, making education more accessible and effective. AI can enhance security systems, helping to prevent cyberattacks, detect threats, and improve public safety, AI has also the potential to boost economic growth by increasing productivity in various industries and creating new job opportunities in AI-related fields, and AI can aid in space exploration, enabling autonomous spacecraft and rovers to explore distant planets and moons.

However, to realize these potential benefits, we must address the challenges and concerns associated with AI. This includes developing robust ethical frameworks, ensuring transparency, and addressing job displacement through reskilling and upskilling programs. Artificial Intelligence is a transformative technology that has the potential to bring about significant advancements in various aspects of our lives. While some people may harbor fears and doubts about AI, it is essential to remember that similar concerns have arisen with past technological innovations, which ultimately became integral to our daily routines.

By educating ourselves, engaging in open dialogue, and establishing ethical guidelines, we can better understand and embrace AI. The future of AI holds great promise, and with responsible development and collaboration, we can harness its power to address global challenges and improve the quality of life for people all over the world. Embracing AI as a tool for positive change is the key to realizing its full potential.

SOCIAL HARDWARE

Writing about "social hardware" involves exploring the foundational elements in a society that facilitate the full realization of individual potential, especially among exceptionally talented individuals. This concept intertwines with how societal structures, resources, and attitudes can nurture and enhance the "intellectual software" of individuals, leading to significant inventions and innovations.

In the context of America, the narrative can emphasize how the country's unique blend of resources, freedoms, educational systems, and cultural attitudes towards innovation and entrepreneurship has historically enabled individuals to achieve remarkable feats of invention and innovation. This has not only transformed the American landscape but has also had a profound impact on the world.

The Birthplace of Innovation

In a world where individual talents shine as beacons of innovation, the concept of "social hardware" emerges as a pivotal element. It refers to the societal and structural framework that cultivates and nurtures individual intellectual prowess, particularly among the highly talented. The essence of social hardware is exploring its impact on nurturing individual "intellectual software" and emphasizes how America, as a crucible of innovation, has leveraged this concept to foster talents that have significantly bettered the global landscape.

Social hardware encompasses the collective elements of infrastructure, education, legal and economic systems, cultural values, and community support that form the backbone of a society. It is akin to the operating system that runs the complex machinery of a nation, setting the groundwork upon which individual talents can thrive and innovate. In the context of nurturing intellectual prowess, social hardware acts as a catalyst, enabling the full potential of an individual's intellectual software - their knowledge, creativity, and problem-solving skills.

America's story of innovation is deeply intertwined with its evolving social hardware. From the establishment of its foundational freedoms and rights to the creation of a robust educational system and the encouragement of a risk-taking entrepreneurial spirit, America's social hardware has been uniquely conducive to fostering innovation. The American experiment began with a profound emphasis on individual rights and freedoms. The founding fathers enshrined these principles in the U.S. Constitution, providing the freedom necessary for individuals to explore their talents and ideas without fear of persecution. This commitment to liberty created a fertile ground for risk-takers and pioneers who were willing to push boundaries and challenge the status quo.

Another cornerstone of America's social hardware is its educational system. The establishment of public schools and the promotion of universal education from an early age allowed a broad spectrum of individuals to access knowledge and develop their intellectual software. Land-grant universities, like the Morrill Act of 1862, further expanded educational opportunities by making higher education accessible to a larger portion of the population. This investment in education laid the groundwork for future innovation.

America's legal system, particularly its patent system, has played a pivotal role in nurturing innovation. Patents provide inventors with legal protections and exclusive rights to their inventions, giving them the incentive to invest time and resources in creating new technologies. This has led to a culture of innovation and entrepreneurship, where inventors are encouraged to transform their ideas into reality, knowing that their work will be protected. Beyond infrastructure and legal mechanisms, America's cultural attitudes towards innovation have been instrumental in fostering talent. The "can-do" spirit, the celebration of risk-taking, and the acceptance of failure as a stepping stone to success have all contributed to a climate where innovation thrives. Innovators like Thomas Edison and the Wright brothers faced numerous failures before achieving their breakthroughs, but they persevered due to a society that encouraged resilience.

Contrasting America's social hardware with that of other nations offers a clearer understanding of its uniqueness. While other countries may excel in certain aspects, such as education or infrastructure, America's holistic approach in combining these elements with a high degree of freedom, flexibility, and support for innovation sets it apart. This unique blend has made America a fertile ground for individual talents to flourish and actualize their potential.

Despite its successes, America's social hardware faces contemporary challenges. The rapid evolution of the economic landscape, increasing global competition, and the need to strike a balance between regulation and freedom are critical issues. Addressing these challenges is essential for sustaining an environment that continues to foster individual intellectual software development. The advent of the digital age and the globalization of innovation have reshaped the landscape. While America still possesses a wealth of resources and opportunities, it must adapt to ensure that its social hardware remains agile and responsive to the changing needs of a fast-paced world.

Global Impact of American Innovations

The innovations birthed from the American social hardware have had a profound and far-reaching global impact. From technological advancements that have reshaped industries and economies to medical breakthroughs that have saved millions of lives, the ripple effects of American innovations are ubiquitous. The development of Silicon Valley, for example, has become synonymous with technological innovation. Companies like Apple, Google, and Facebook have not only transformed industries but have also changed the way people around the world communicate, access information, and conduct business. In the field of healthcare, American innovations in biotechnology and pharmaceuticals have led to breakthrough treatments and life-saving drugs. The development of vaccines, underscores the global significance of American contributions to public health.

The concept of social hardware is crucial in understanding the dynamics of innovation and intellectual development. America's experience illustrates how a well-structured social hardware can serve as a fertile ground for nurturing individual intellectual software. This synergy between societal support structures and individual talent is not just the cornerstone of American innovation but also a blueprint for other nations aspiring to ignite their own eras of technological and intellectual advancement. In a world increasingly reliant on innovative solutions to complex challenges, the nurturing of such individual talents through robust social hardware is more important than ever. The concept of social hardware can also serve as a testament to the power of a supportive societal framework in unlocking human potential.

"Social hardware" represents the foundational elements in a society that facilitate the full realization of individual potential, particularly among exceptionally talented individuals. America's unique blend of resources, freedoms, educational systems, and cultural attitudes towards innovation and entrepreneurship has historically enabled individuals to achieve remarkable feats of invention and innovation.

As we navigate the challenges of the 21st century, it is essential to recognize the critical role that social hardware plays in fostering innovation and nurturing intellectual software. By learning from America's experience and adapting social hardware to meet the evolving needs of the world, nations can empower their own talents to shape the future and address the complex challenges that lie ahead. The legacy of social hardware is a testament to the potential of individuals and the power of a supportive society to drive progress and change the world.

AMERICAN EXCEPTIONALISM – OR EXTRAORDINARISM?

American Exceptionalism, a term that has sparked debates and discussions for decades, is more than just a belief in the nation's superiority. At its core, it encapsulates a unique blend of diversity, inclusion, and individualism, setting the United States apart from other nations. Unlike some countries where ethnic homogeneity or a sense of ethnic superiority prevails, America's strength lies in its rich tapestry of varied backgrounds, races, and cultures. In this melting pot, every individual, regardless of their origin, is a thread that contributes to the larger American narrative.

Let's delve into the multifaceted nature of American Exceptionalism, exploring its historical roots and how it is shaped by the nation's diversity and commitment to individual freedoms. It argues that American Exceptionalism is not about a homogeneous sense of superiority but about a system that cherishes freedom, democracy, and the potential for individual achievement. This system has not only fostered a culture where individual dreams can flourish but has also become a magnet for geniuses and dreamers worldwide. Through this exploration, we can better understand how the American ethos of inclusion and individualism makes the United States an exceptional place where possibilities are as vast as the sky itself.

Individualism in America and the American Dream

At the heart of American culture and ethos lies a deeply rooted sense of individualism, a value that has shaped the nation's identity and fostered the concept of the American Dream. This notion of individualism, characterized by a strong emphasis on personal freedom and self-reliance, is not just a cultural trait but a cornerstone of American society, influencing everything from political ideologies to everyday life.

The roots of American individualism can be traced back to the country's founding principles. The Declaration of Independence and the Constitution reflect an understanding of freedom and individual rights as fundamental to the American way of life. This emphasis on individual liberty and the pursuit of happiness has been a guiding force in American history, driving innovation, exploration, and a spirit of entrepreneurship.

The American Dream, often intertwined with individualism, is a multifaceted concept. It's the belief that anyone, regardless of their background or social class, can achieve success and upward mobility through hard work and determination. This dream has been a magnet for millions worldwide, attracting immigrants seeking a better life and opportunities unavailable in their homelands.

Throughout the 19th and 20th centuries, the American Dream was often associated with homeownership, a stable job, and a better future for one's children. The post-World War II era, especially, saw a boom in economic prosperity, making this dream more accessible to a broader segment of society. In the realm of entrepreneurship and business, individualism has been a driving force. The U.S. is known for its entrepreneurial spirit, with stories of individuals starting from nothing and building successful businesses becoming a key part of American folklore. Icons like Henry Ford, Steve Jobs, and Oprah Winfrey, among others, embody this ethos, showcasing how innovation, risk-taking, and hard work can lead to extraordinary success.

The 21st century has seen a reevaluation of the American Dream, with newer generations interpreting it in the context of a changing world. Today, the dream might be less about material wealth and more about life satisfaction, personal fulfillment, and social mobility. The digital age, along with the rise of gig economies and technological advancements, has also reshaped what individual success looks like.

Moreover, individualism in America is not just about economic success; it's also about personal freedom and self-expression. This aspect is evident in the country's vibrant cultural scene, where individual creativity and expression are highly valued. From the diverse music genres that have emerged from the U.S. to the array of literary and artistic movements, American culture is a testament to the value placed on individual expression and innovation.

Individualism and the American Dream are fundamental to understanding America and its place in the world. While these concepts have evolved and faced challenges, they continue to inspire and drive the nation. They are not just ideals but lived experiences that reflect the complexities and dynamism of American society. As America moves forward, these principles will undoubtedly continue to shape its future, adapting to new realities and continuing to offer hope and inspiration to many.

American System: A Beacon of Freedom and Democracy

The American system, marked by its robust commitment to freedom and democracy, stands as a beacon to many around the world. Unique in its composition and operation, this system has historically attracted brilliant minds seeking to realize their aspirations in a land where possibilities seem endless. At the heart of this allure is a distinct blend of liberties, democratic principles, and an ethos of opportunity that differentiates the United States from other nations.

Diversity in America is more than just a demographic feature; it's the bedrock of societal strength and cultural richness. The nation's history as a melting pot for various ethnicities, races, and cultures has fostered an inclusive ethos. This diversity has encouraged a multitude of perspectives and ideas, contributing to innovation and progress in various fields. It symbolizes the reality that different backgrounds and viewpoints are not only accepted but are integral to the nation's fabric.

Individualism, another cornerstone of American Exceptionalism, empowers personal freedom and self-reliance. It encourages citizens to pursue their unique paths, fostering a culture where innovation and entrepreneurship are highly valued. This ethos has given rise to a society where success is not predetermined by one's background but can be achieved through hard work and determination. The American Dream, an embodiment of this individualism, continues to inspire both natives and immigrants to strive for personal and professional fulfillment.

The democratic system in the U.S., characterized by its commitment to individual rights and liberties, stands as a model of governance. It provides a framework where diversity and individualism can thrive. The balance of power, free elections, and the rule of law are not just political ideals but practical tools that support a society where anything is possible.

While many Americans may not consciously feel "exceptional," the collective system and values in place create an environment that is indeed extraordinary. This exceptionalism is not without its challenges, and the journey towards a more inclusive, equitable society is ongoing. The aspiration is for a world where this model of diversity, individualism, and democracy is not an exception but a norm, creating a global environment where possibilities are limitless, and success is accessible to all, regardless of race, ethnicity, religion, or culture. This vision of a universally exceptional world represents the next frontier of global progress, inspired by the American example.

ABOUT THE AUTHOR

In the process of writing this book, my journey through the landscape of American inventions and innovations over the last 150 years has been nothing short of enlightening. It has been a profound exploration, not just of the tangible advancements that have shaped our modern world, but of the spirit of creativity and resilience that defines the United States of America. This exploration has deepened my appreciation for America's role as a crucible of innovation, where ideas are nurtured into inventions that have not only propelled America forward but have also made significant contributions to the global community.

As you turn the pages of this book, I hope it serves as a bridge to a broader understanding, perception, and perspective of America. By delving into the narratives of these groundbreaking achievements, my aim is for you, to gain insight into the multifaceted nature of American progress. It is my hope that this exploration will not only enlighten you about the remarkable ingenuity embedded in America's history but will also inspire a deeper reflection on how these inventions continue to influence our lives and the world at large.

Let this book be a starting point for further exploration and curiosity about the ways in which American innovation continues to shape our future. May it broaden your horizons and encourage you to appreciate the interconnectedness of our global community, united by the shared pursuit of progress and the common good. Through this journey, may you find renewed respect for the power of human creativity and the endless possibilities it holds for bettering our world.

www.AmericaFromZipperToSpaceX.com